MediaSpeak

Second Edition

Strategy. Sound-Bites. Spin.

The plain-talking guide to issues, reputation and message management

Ian Taylor

Published in Canada by
MediaSpeak Press

National Library of Canada Cataloguing in Publication

Taylor, Ian, 1950-
MediaSpeak : strategy, sound-bites, spin / Ian Taylor. —
2nd ed.

Includes index.
ISBN 0-9734472-0-6

1. Public relations. 2. Public speaking. 3. Public
relations—Case studies. I. Title.

HM1221.T387 2004 659.2 C2004-900929-X

Table of Contents

Part One: Planning your MediaSpeak strategy

Chapter One
Bimbos, banks and spin 1

What this book's about 1

We're all spin doctors 2

What is public relations? 4

What is spin? Spin is word choice. 5

Case Study: Caring about ambulance service — Crisis scenario 6

Spin doctors 7

You and the news media 10

Dealing with the news media can be a bit like going fishing — A parable 11

Preparing the answers 12

A reputation for caring 14

A reputation for showing that you care 16

You can always tell a lawyer 17

Case Study: Caring about justice — Lawyer Clayton Ruby 17

There's strategy in how you say it 18

The messages of change management 19

Shooting the messenger 20

What was Clinton's strategy? 22

What was Clinton's method of delivery? 23

Waving at trees 24

Chapter Two
Packaging the message 27

It's about packaging 27

Preparing the package 28

If it bleeds, it leads 30

What makes you newsworthy? 30

Your emotional quotient 33

Case Study: Caring about squeegee kids 33

Write a letter to the editor — today! 35

How big is your budget? 36

Case Study: Caring for cleanliness — Sunlight Soap 36

News is what's new — even when it's old 38

A reputation for speaking well 39

Who's your boss? 40

Chapter Three
The message will set you free 43

Introduction to *MediaSpeak* 43

Is the medium really the message? 44

"Welcome to the evening news. I'm Ian Taylor." 45

Where does TV get its ideas? 47

How Oprah selects her guests 47

Copycat journalism 48

A story takes on a life of its own 49

The effect of the global village 50

Some questions to ask a reporter before agreeing to an interview 51

What not to ask a reporter 52

Case Study: Caring about safety —
Flying truck wheels in Ontario 52

Media requirements are based on
space and time 54

How long is my media line? How long is
the interview? 55

"What we have here
is a failure to communicate." 56

A reputation for word-crafting 58

The credibility factor 60

Comforting the afflicted 61

Fill in the blanks 62

Chapter Four
The skills to succeed **63**

A reputation for plain talk 63

"First of all, let me say…" 64

Tips to improve your pick-up rate 67

Good writing prevents misquotes 68

Learn to write like you speak
and stop speaking like you write 69

A reputation for speaking out 69

A reputation for looking right 71

Body language: Neurolinguistics 71

Message dressing 73

A reputation for sincerity — you'll have it made 76

A reputation for understanding differences 78

A reputation for good intentions 78

How to manage a scrum 80

A reputation for avoiding becoming the target 80

A reputation for getting more resources 82

Chapter Five
Be bold, or be toast **85**

With *MediaSpeak,* it's the answers, stupid — no matter what you're asked 85

Case Study: Caring about your employees — Hospital layoffs 88

A reputation for finding opportunities 90

Start with one word or phrase 91

Direct and indirect quotes 92

A reputation for openness 92

Self-deprecation 93

Part Two: Writing and Delivering MediaSpeak

Chapter Six
Pillars 97

Highlights 97

A reputation for getting to the point 99

How PR people overcome their reputation 101

Case Study: Still caring about safety —
Photo radar 102

Case Study: Caring about customers —
Airport taxis 103

Case Study: Caring about the environment —
Aircraft noise and neighbors 105

Chapter Seven
Reputation words and the “not” word 109

Negative questioning 110

“Not” is a dangerous word 111

Opportunity words from the “reporter from hell” 113

Planting the pillar foundations 117

Issue-defining bridging phrases 117

Here’s another way to get started 120

Case Study: Caring about business —
Escalator safety in shopping center 121

Speculations 123

Case Study: Caring about your
professionalism — Highway tolls 124

Buying time 126

Super-controlling pillars 127

Chapter Eight
Never say "No comment" 129

Your job is to inform, describe, explain 129

There are lots of reasons never to say "No comment" 130

Case Study: Caring about immigration — Deporting the maid 131

"No comment" alternatives — When the first word is the last 131

A reputation for being ready for anything 134

Favorite questions from the "reporter from hell" 136

Ask yourself the tough questions first 138

Ask yourself open-style questions 139

Is it on your business card? 141

Define your role in the issue 142

Samples of role-defining pillars 142

Pillars that frame or re-frame the question 144

Chapter Nine
Concise, direct and interesting 145

Can you start with two-point or paradox pillars? 145

Taylor's top ten reasons to use pillars with three's 146

A reputation for teamwork 147

Strategic tips for using three's 147

Factors and factoids 148

Power pillars 152

The click 153

A reputation for order 153

Chapter Ten
Supports — Know when to hold 'em & when to fold 'em 155

Highlights 155

Keep them short, sweet and to the point 157

Reporters love clichés — even though they deny it 159

MediaSpeak from marketing language 160

My Aunt Betty had quite the reputation 161

Malapropisms and old favorites 163

A reputation for "humbility" 163

Euphemisms 164

Analogies are like tools 165

Control supports 166

Bridges 167

Baiting supports 168

Phrases to avoid in an interview 168

The power of repetition 170

Helpful phrases to slow down the charging reporter 171

Handling the ambush allegation 172

The pause 173

Chapter Eleven
Sparklers — Displaying your excellence 175

Why sparklers pay off 176

Rattle off some numbers 177

Use teaching aids 178

Anecdotal sparklers 178

Case Study: Caring about agricultural land —
Farmer Brown's daughter 178

Unique features sparklers 179

The sparkler as parable 180

The "chapter and verse" sparkler 181

The "for instance" 181

When is it OK to cry? 184

The sermon sparklers 185

Context, background and history 186

A reputation for story-telling 187

Case Study: Caring about history —
Open line opportunity 188

MEDIA LINES — A summary 189

A Spin Doctor's Lament 190

Appendix

Free Article: *Write Better News Releases* 193

Biography: Ian Taylor 198

Part One

Planning your MediaSpeak strategy

Chapter One

Bimbos, banks and spin

What this book's about

The great spin doctor Anonymous once said that character is what we are; reputation is what people think we are.

We can't know what people think about us, however, until we know what they say and do about us, and today, public opinion is heavily influenced by the messages in the news media. In order to deal with the news media messengers, we need to manage our message — the strategy, the choice of words and the delivery of those words. Strategy. Sound-bites. Spin.

Consider the messages in these headlines:

"Thousands to lose jobs in bank mergers"

"Bank mergers good for the economy, Chamber of Commerce told"

"Personal bankruptcies at record high while bankers celebrate"

"Bank exec. denies gouging consumers on bank fees"

There's no corporation more careful of its reputation than a bank. Unlike politicians, however, banks do not seek traditional popularity, they seek customers — big ones. My father used to say that if you owe the bank $10,000, you're a pain in the neck (he didn't say neck, of course) but if you owe them a million, you've got yourself a partner and a best friend.

Canadian Banks have announced record profits in recent years. This is **news "with legs"** — the story lasts a long time and has lots of opportunities for reporters to cover issues in many ways. Banks are a big news story and everyone, it seems, has an opinion about them. That public opinion forms the basis of their reputation. As banks seek to merge, they've become even more conscious of their messages, who delivers them and how.

Welcome to *MediaSpeak,* where we'll introduce you to the skills and techniques employed by the experts to manage their messages in almost any medium or communications encounter. Your *MediaSpeak* skills will help you deal with customer service issues, marketing, presentation skills and any area where you communicate with others.

In this book, you'll learn how to develop clear strategies to position your message in the public interest while displaying your professionalism, along with that of your organization.

You'll learn to "package" your messages using easy-to-understand methods for composing your core communications statements.

You'll learn the art and science of the sound-bite.

You'll learn how to answer even the most hostile or aggressive questions from the "reporter from hell." And you'll learn to adapt those answers to a host of communications encounters.

We're all spin doctors

In the future, everyone will be famous for 5.2 seconds! That's the average length of a "sound-bite" on the evening news.

You have to earn a living. If your job involves communicating with others (and whose job doesn't?) that may mean mastering the art and science of the sound-bite. In the very near future, more of our computers will be equipped with screen-mounted video-cameras. We'll move from e-mail to v-mail. Our on-camera skills will be important in many jobs where they never were before. Some financial institutions will require advances in video reporting and they'll need to learn from the experts.

Marshall McLuhan said that the medium is the message. **I believe the message is the message**, and that good messages will stand up to adaptation in all types of media formats, whether for broadcast, print or verbally delivered.

If your job depends on how well you communicate, you can learn a lot from politicians like former US President Bill Clinton. Through Whitewater, Bimbogate and other crises, Clinton managed his message in order to meet his goal of keeping his job.

Most successful politicians follow a message-management formula in order to prevent issues from becoming crises. That's what this book is all about — developing communications strategies, preparing and sticking to a planned message and delivering that message through the news media and other communications formats.

Those formats may include written publications, customer service encounters, marketing messages, web-page information and other media, but we focus on the mainstream news media who, very quickly, can damage or build a reputation.

If your job is on the line when you communicate with the public, you'll need to learn a few skills about managing your message with *MediaSpeak*. At the same time, with those skills you'll learn to improve, enhance, or at least avoid damaging a reputation.

According to the CBS-TV program *60 Minutes,* Bill Clinton's staff regularly conducted "message-testing" — using public opinion polls and/or focus-groups. The purpose? To tell people what they often want to hear, or to deliver *messages* that will be popular.

Like Clinton, the bank bosses want to keep their jobs, but to do so, must protect their shareholders and satisfy their boards of directors. Banks set clear communications strategies, stick to a message and deliver it in highly contained formats. They even equip their tellers with scripted messages to handle customer complaints.

I've trained a few politicians and a few corporate executives in the financial sector. They do not form the bulk of my business, however. I'm more likely to be training someone from the public, transportation or healthcare sector. My students are involved in emergency and crisis management. I tend to train middle and senior management, public administrators and communications staff. This book deals with the kind of issues they're likely to face and they can learn a lot from Bill Clinton and the banks.

What is public relations?

I believe that public relations is really about the management of change — either by changing what you're doing to foster public support and/or by changing what you're saying to foster public understanding, and hence, support.

PR people have a host of names for their jobs — public affairs, communications, reputation management consultants. They work in all aspects of organizations and often have responsibilities for fundraising, shareholder relations and project management. They're often the first blamed when things go wrong, the last to get credit when things go well. They're only part of the public relations function in an organization.

I believe that the duties and responsibilities of public relations are part of everyone's job, or at least everyone who deals with the public.

Here's the official, three-part definition from the Canadian Public Relations Society:

> Public relations is the management function which
> evaluates public attitudes, identifies the policies and
> procedures of an organization with the public interest,
> and plans and executes a program of action to earn public
> understanding and acceptance.

What is spin? Spin is word choice.

As the bishop said to the showgirl, life is about choices. To manage the message in a media interview or elsewhere, you must choose the words based on a planned strategy and then deliver those words effectively to your audiences.

Let's look at an example that illustrates what is called "spin."

A news editor or business reporter may choose to cover banking issues in several ways, applying a different approach, "angle" or spin to the same basic news item. Some possible headlines:

"Banks continue to flourish"

"Bank executives reap record pay checks"

"Service charges account for huge profit"

"Small businesses still ignored by big banks"

And as the story is played out, new or different spin is often added as other organizations become involved:

"Pensioners starve as bankers dine on caviar"

"Bank's annual meeting to be hot affair for shareholders"

"Hard-hit charities call for share of bank profits"

"Bank president marries trophy wife, career in jeopardy"

"Containing the message" involves limiting and highly training a small, select number of official bank spokespersons available to the news media, yet equipping every part of their communications operation with consistent messages, down to the bank tellers who work on the front lines of customer service.

One thing that politicians and the banks understand is that they **cannot manage the news media**, but they **can, and must, manage their message**.

Even famous media moguls like Ted Turner, Rupert Murdoch and Conrad Black can not control what their competition says about them. They can, and do, however, control what they say about themselves — including how, when and where they say it.

This book contains dozens of communication situations, sample statements and quotable quotes based on a wealth of experience in media relations. I wrote my first letter to a newspaper editor when I was 16. My first live interview was on CBC radio when I was 23. I've been interviewed by reporters thousands of times in the past 30 years. I had lots of chances to make almost every mistake possible. It's my hope that you'll learn from my mistakes so that you won't have to make them yourself.

Case Study: Caring about ambulance service — Crisis scenario

A heart attack victim dies while waiting for an ambulance. The ambulance service has been reduced due to budget cuts and government re-organization. The resulting news headlines or spin on the story:

"Union calls for enquiry into ambulance crisis"
"Heart attacks — what to do until help arrives"
"Seven minutes of agony — a family's painful wait"
"Drug company calls for approval of home heart medicine"
"Hospital chief calls for task-force on emergency services"
"Ambulance crew at donut shop during emergency call"
"Death blamed on government cutbacks"
"Family calls for changes in ambulance service"

Picture yourself as head of the ambulance service, or as city councillor for the area or president of the ambulance workers' union. As head of the ambulance service, you'll want to be very open about the issue, adopting what's often called the "victim of circumstances" strategy, without using the word "victim."

While you might not choose to use the word "crisis," you might choose to describe the situation as "very serious." This makes you a spin doctor.

In each of these headlines, a different spin is being applied — by external forces often outside your control; by media decisions at every level of the news organization, often outside the role of front-line reporters; and by the public's insatiable appetite for info-tainment.

News is a business with lots of competition and ambulance services impact everyone. Each of the groups or individuals involved in the headlines above (including the reporters and editors) are involved in spin doctoring.

Education is much in the news today. Picture yourself as a concerned parent, a teacher or a publisher of textbooks. The government introduces sweeping changes to the education system, announcing its commitment to improve the quality of education. Each participant in the issue might bring to the table a different spin or angle, based on their level of interest, their stake in the issue, and what they hope to change or protect.

Teachers, unions, school boards, parent groups and lots of "instant experts and pundits" line up to criticize or support the changes, forecasting various results on future generations of students. Who's right? Who's believed? Whose reputation is at stake? Who's best managing their strategy, script and delivery? Spin becomes highly competitive, between a host of agencies and voices.

Spin doctors

In the back rooms, and out of sight of the public, toil the professional and not-so-professional word-crafters *whose job it is to find media opportunities and avoid pitfalls!* People sometimes called **spin doctors**. I have never been ashamed to call myself one.

Toronto Star reporter Rob Salem, in a Feb. 2, 1997 story, wrote:

> "Flak, toot, shill, spin doctor, bumf jockey, tub-thumper, promulgator of propaganda, highly placed source, Velcro buddy, new best friend… Publicists publicize. They fax, they phone, they issue endless [news] releases. They adamantly refuse to take no for an answer. They arrange interviews and set up photo opportunities, organize and cruise events and administer guest lists."

Salem interviewed a number of publicists in the entertainment business, who listed among their philosophies: "**Determination, perseverance, responsibility, a sense of humor, professionalism, flexibility, timing and luck.**" One of them told the Star, "It's not brain surgery," while another admitted, "You're only as good as your last gig." I happen to believe you're only as good as your next gig.

Entertainment **publicists** work in a specialized field of media relations. On the release of a new movie or recording, the stars will be available for media interviews that can be stage-managed like factory assembly lines. Reporters are clamoring for the opportunity to talk to screen legends. Yet try to get coverage for your neighborhood little theater and you'll hear how over-worked and under-staffed the newsroom is and how they'll take a look at your material but can't promise anything.

Actor Dustin Hoffman is reported to have given 60 interviews in one day as part of the release of the movie *Wag the Dog*— about spin doctors in Washington. A very scary movie, and rather prophetic, by the way.

Do you suppose Hoffman changed his message much in those 60 interviews? Do you suppose many of the questions were original or creative? Do you suppose he might have been carefully scripted, just like the movie? (Notice the purposeful repetition in the three "framing" questions here.)

Do you suppose that Hoffman altered his answers very much during all those interviews? Do you suppose he suddenly decided to let loose and start talking about his true feelings about his co-workers? That he was suddenly smitten by an urge to "wing it" or ad-lib about his personal opinions on politics? Chances are, he was coached and advised by a spin doctor at least some time in his career, or else he learned his lessons by making mistakes.

On the other side of the coin are those spin doctors or public relations people **whose job is to keep their clients out of the media.** To prevent news coverage or deflect it away from their clients or their clients' issues. Or to clean up a mess by clarifying their clients' statements. Or to contain the story when it can't be fully controlled. Or, to go before the news media and deliver a patently unbelievable falsehood, or else look for a new job. I've had to do it and it's not always fun cleaning up behind the elephants after the circus has left town.

In the PR business, there are old spokespersons and there are bold spokespersons. There are very few of us old, bold spokespersons.

It's an interesting situation when dealing with reporters. It can be almost impossible to get media attention — no matter how good the news release or how important the story seems to you. But once you've got media attention, it can be nearly impossible to avoid it. One reporter colleague told me he can get anyone's phone number in less than 10 minutes, and that was before the Internet.

Media attention is a lot like greatness: some are born with media attention, some achieve media attention, and others have media attention thrust upon them.

You and the news media

You will be of interest to the news media for three main reasons or a combination of them:

- ☞ **Your job, position or role relates to a topic or issue that is in the news or is potentially newsworthy.**

- ☞ **You have a particular interest in an issue and you symbolize or represent a certain aspect of the issue, no matter how large or how small that aspect is.**

- ☞ **Your expertise, knowledge or experience relates to the topic or issue, and you can therefore contribute to public awareness or changing public awareness.**

You'll make the reporter's job easier and you'll communicate better if you know:

- ☞ **how the media works and how reporters gather information.**

- ☞ **how to "package" your message in a way that meets the media's needs. This includes the strategizing, writing, and delivery of your message.**

- ☞ **how to make yourself available and visible to reporters.**

- ☞ **how to refer the media elsewhere or say very little.**

- ☞ **how the experts take control of media opportunities.**

- ☞ **how to avoid potential problems, reporter traps and unplanned statements.**

Dealing with the news media can be a bit like going fishing — A parable

There are certain times of the day or the year when the fish are out biting because food is scarce and they're hungry for some new morsel. At other times, the fish are full, lazy and not the least hungry.

To catch fish, you have to go where the fish are. Sometimes that means following the other fishers and sometimes it means finding an all-new spot. There are dozens of new TV channels, new magazines, new Internet opportunities. Do you cast your line in a new corner of the lake or in the old, familiar places? Or do you throw a wide net hoping for anything? Are there holes in your net?

Do you use all the new, bold and innovative techniques of message management, or stick to the old ways because that's how you've always done it? Do you still, for instance, sit down for a news conference or TV interview when it's more effective to talk standing up?

You have to use the right bait for the right fish. The bait is entirely based on the quality of your hook — your early message. And if it's not working, change it by re-writing or re-strategizing it.

Watch what the other fishers are using to catch fish. If they've found a line that works very well, chances are you can learn a thing or two. At the same time, if others notice your success, they'll quickly change their methods, too. Soon, the water is chock full of lines and the fish have moved off to another area.

You want to avoid being eaten by a whale while you're looking for trout. You might consider yourself really lucky at first, until you spend 30 days in the whale's stomach, only to be thrown onto the beach naked, at the end of the experience.

You might spot a school of fish that looks like easy pickings. You think you can bluff your way into the headlines. Then suddenly, the sharks emerge at the very first smell of blood in the water. They've done their homework this time. All the rules change.

Even the tiny, seemingly friendly fish can suddenly turn nasty. A feeding frenzy occurs and it doesn't end until there's little left alive in the water. The fishing trip can be very short indeed.

You can cast out your line, but remember, there's a hook on the end. If the hook is not catching fish, try a new hook, or go fish somewhere else. You can't always cut the line and disappear off the water without an expensive rescue helicopter standing by. And wear a life jacket. It's called the truth.

Preparing the answers

Canadians who followed the 1997 news on the ill-fated Somalia Enquiry will be familiar with the terms used by the public affairs office of Canada's Department of Defence — "media response lines" and "responses to queries." Or, in military-speak, MRL's and RTQ's. Dozens of staff-members were found to have been involved in watering down or altering files containing media messages after an access to information request from a CBC radio reporter.

News reports after the death of Princess Diana suggested that Queen Elizabeth, in an attempt to improve her reputation, had suddenly hired a spin doctor. This is a bald-faced lie. Princes, popes and potentates have always had spin doctors — they just work under a wide variety of job titles — speech writers, advisors, counselors. Spin doctors have gone to great lengths to spin themselves new titles, the latest — consultants in reputation management.

Those with the most to hide are the ones most protesting the word "spin." Any PR people who say they're not spin doctors are lying or they're not doing their jobs — to influence public opinion.

In Britain, Prime Minister Tony Blair's spin doctors refer to the scripting process as being "**on message**." Blair is reported to have been highly influenced by Bill Clinton's staff, some of whom were shipped off to London to help elect Mr. Blair. And to advise Mrs. Blair.

In Washington, Monica Lewinsky was alleged to have been provided with media "talking points" to use while fending off reporters asking about her affair with Bill Clinton. She did, however, hire a media consultant, which the media would have us believe is a new trend. Spin doctors have been around since the cave artists, whose job, it's been said, was to make certain body parts appear bigger than they actually were.

Some spin doctors refer to media lines as "talking points." Some politicians call them "the party line." All the experts use some variation of media lines to prepare for a media interview or manage their communications strategies, and you can, too.

One of the most important sections of any media line may be the part which instructs: "If asked about this topic, stress this position..., or, if they ask you about this topic or issue, here's the suggested answer." And, in the world of customer service, you'll learn to train your staff so that if the customer complains about this topic, say this.... If I complain about bank charges, I'm sure my bank employee will provide me with a stock response; at least he or she should be trained to do so.

Here's the bonus of this system in front of a TV camera: you'll always be ready with a sound-bite or quotable quote, whether the reporter uses it or not. That's something that can almost always be said about Bill Clinton, or, for that matter, about Greenpeace or many high-profile organizations seeking media attention. It's about more than just the questions. It's about having spontaneous answers ready for almost any question, in any communications or customer service encounter.

Reporters sometimes ask questions that are designed to elicit a certain specific response. There are techniques that help them put words in your mouth — if you fall for the traps.

You can spend much of an interview reacting to a reporter's negative questions, or you can be pro-active. You can start answering questions that haven't even been asked yet. You can volunteer information.

One thing is certain, however. As soon as a reporter gets you looking and acting defensive(ly) you'll start to look guilty. And as soon as you start to **look** guilty, people will think you **are** guilty. As soon as you say, "There's no cover-up," some people will start to think that there is one.

As reported during the Clinton impeachment trial, when someone says it's not about money, you know it's about money; and when they say it's not about sex, you know it's about sex. What a revelation about public opinion!

When you fall for negative questioning, you're part of the problem instead of the solution and you've failed an often basic strategy — **to position yourself as a victim of the situation**. To make it someone or something else's fault.

A reputation for caring

There's an old rule in media relations — if you want to inflame an issue, use inflammatory words, especially to deny. If you want to control an issue, control your words. Use a media line. We'll teach you how to prepare one.

In a later chapter, we'll provide a list of about 400 negative, accusatory words that reporters use to get you to contribute to creating controversy — **simply by repeating them and denying them**. We'll illustrate how reporters plant these words in the question in order to get you to use them in your answer, creating instant sound-bites. By sticking to your media line, you won't become a victim of the questions, but you'll remain in control of the answers.

Media lines already exist in many organizations. They're part of your communications and employment culture. They're found, to some degree, in your annual report, brochures and publications, your customer service language and in form letters. They're likely already a part of your internal communications materials, speeches and marketing messages. If not, they should be.

In many cases, you need not be saying things in a brand new way. Chances are, you've already got perfectly good answers to every possible question that may arise. They're from within the speech you just gave, the report you just tabled or the answers you've used before.

You've probably seen the experts on television — the ones who can handle almost any question with **public-spirited professionalism**. These are people who seem to have **spontaneous answers ready to any questions**, even though they may have used that answer in dozens of other interviews.

Some of these spokespersons have had extensive training, others seem to be naturals. You can learn from their successes and their failures every day, simply by reading the newspaper or monitoring broadcasts differently and by watching for the techniques described in this book.

It's never obvious, but the on-air TV personalities you'll face have a lot of advantages. They've been dressed in a studio and don't have a wrinkle on them. Their hair and makeup have been done for them. Someone has researched most of the questions or is prompting them through an earpiece. And they've been doing their job for a while. You may be experiencing your first time on TV. The reporter has the advantage of the format, but you have the advantage of the answers you bring.

What is most important here is that you may have to learn or re-learn or re-apply some of your most basic communications skills. You may have to change your organization's policies and practices. But more than any other strategy, **you'll need to show that you care** about people, about issues and about your community. You'll place your issues and concerns in the public interest to succeed.

A reputation for showing that you care

The great philosopher Anonymous once said that people don't care what you know until they know that you care.

The essential starting strategy in any communications encounter is to say and to show that you're concerned about others. You must position your issue in the public interest rather than the narrow self-interest of an individual or greedy organization. This is a variation on the old fashioned strategy that what's good for General Motors is good for the USA. Today, it has an all-new focus.

> **As one politician told me, "Never let anyone else out-concern you on an issue."**

In a media interview, just as in a customer service encounter, you'll have to learn some techniques that the experts use to look comfortable and be comforting. In order to look comfortable, you'll have to be comfortable with your message.

You'll have to learn how to listen to reporters' questions so that you can determine very quickly what the question is about rather than be trapped by specific wording. We've had long discussions on this topic with clients who are courtroom lawyers. An edited newspaper interview must be conducted much differently than a case summary.

Advertisements by banks go to great lengths to say that banks care. Unfortunately, they do a poor job of showing it, as any business-owner can tell you. One series of advertisements talks about bank staff visiting customers at home, on Saturdays and weekends, on their coffee breaks and after regular hours. Try calling your bank and enquiring about this service and see if you can do business with them at your convenience.

You can always tell a lawyer...

It's one thing to do battle and communicate in a court of law. It is a unique format in which the entire packaged message is generally taken into account. It's something else to do battle in the court of public opinion, where you're only as good as your weakest sound-bite.

Case Study: Caring about justice — Lawyer Clayton Ruby

The story is told of Toronto lawyer Clayton Ruby being asked if he would agree to an interview on the steps of a courthouse at the end of a day of trial. Ruby asked the reporter how many seconds of air-time she might be seeking.

When told that 15 seconds might be used on-air, Ruby delivered a 14.3 second statement and then thanked the reporter for her interest.

When the surviving Dionne quintuplets took the Ontario government to court, they chose Ruby as their lawyer and he served them well.

Not every lawyer has Ruby's skill in the court of public opinion. Mastering the sound-bite can be very good for business, lawyers find.

Some of the best spokespersons in the world are lawyers. Many politicians learn their early skills in the practice of law. Yet when a major charity, CARE Canada, faced a crisis over allegations of fund-raising irregularities for Somalia, it was reported that their lawyer had advised them to say "No comment" to the CBC national news and the issue would go away. It didn't, of course.

Lawyers without media experience should stick to "lawyering" until they learn the ropes. I've trained hundreds of lawyers, and I admire some of them. I even have friends who are lawyers. I'm not a lawyer, but I charge like one.

Your interview strategies often include avoiding traps in the questions, and re-defining the issues from the way they're presently defined by the reporter or by your critics.

Your strategy may be to raise new questions and deflect the "blame" or responsibility carefully without being seen to be pointing fingers. You may wish to raise new issues that haven't been asked. Your goal may be to say, without ever using these words — "It's not my (client's) fault." Welcome to the world of spin.

There's strategy in how you say it

Studies show that our words account for only a part of the total communications package. Our appearance, voice, delivery and style all contribute to the final result. An actor must learn to play a role convincingly. It worked for Ronald Reagan.

The experts offer a wide range of views on the impact of words versus the impact of delivery. Some suggest that words account for only a small portion of the message impact. Try watching the evening TV news with the sound turned off. You'll still get much of the message from non-verbal communication. Two actors can deliver the same set of lines and the result can be vastly different for each one.

The same can apply if two vice presidents of your company deliver the same media message — one of them may appear sincere, the other pompous or out of touch. One may come across as part of the problem while another comes across as part of the solution. **Part of your strategy is to choose the right spokesperson, regardless of their title or role.**

It's never enough to say that you care. Managing the message is about much more than what you say, it's about what you do and therefore, what you say that you're doing. And as a result, it's about what gets said in the news media based on your influencing the message.

Susan RoAne says it well in her handy book, *How to Work a Room,* (Shapolsky Books, 1988). "University educator Ernie Baumgarten told a seminar group, '**Behavior that doesn't support words, actually subverts them.**' If what we *do* doesn't support what we *say,* we're worse off than if we hadn't said anything at all."

In other words, we have to walk the walk if we talk the talk. We have to practice what management guru Tom Peters calls "**management by walking around.**" I believe that message management is not about declaring one's excellence, it's about displaying it.

Whether you're concerned with abuses of child labor, like the activist, Craig Keilburger, or whether you're a clothing manufacturer explaining your employment and contracting practices around the globe, you'll need strong communications skills to deliver your message effectively through the news media.

When sporting apparel company Nike was faced with the issue of third-world child labor, they had a crisis on their hands. Groups were calling for boycotts. Nike changed their policies, with tons of free publicity as a result. As one of my first bosses, Mac Erb, used to say, "You can't do dumb things and get good ink. Doing smart things doesn't assure you of good ink, but it helps."

The messages of change management

Even the bankers' marketing messages agree: "**The times they are a-changing.**" According to some experts, consumers are becoming more "empowered" with greater communications skills. Read the book, *Sex in the Snow: Canadian Social Values at the End of the Millennium,* from Penguin Canada, by Michael Adams.

In an excerpt from the book, Adams says:

> "…the opinions and attitudes of the general population are more important than ever before, for the simple reason that the viewpoint and the expertise of our putative elites matter less and less.
>
> I believe that by the year 2020, the institutions the boomers fought to reform will have much less significance for Generations X, Y and Z. Organized religion, institutions like universities, the professions and yes, even the nation-state — all will be much less relevant.

> I believe that the media-rich environment in which we live is making it easier for people to construct for themselves sets of values that are not limited by personal demographic considerations."

Are you one of those people challenging existing institutions or are you a representative of one of them? Neither of you will have automatic credibility. In either case you know that **the public is as mad as hell and they aren't going to take it any more**. Brand loyalty is becoming a thing of the past. Even church growth experts point out that today's "consumer" of church is likely to shop around and jump from church to church for reasons that may have to do with parking or the quality of music or programming.

Shooting the messenger

A CBC-TV documentary entitled *Dawn of the Eye* claims that television is the most powerful communications medium ever invented. Yet, as new forms of inter-active media are being developed through computerization, the producers of the documentary may really be talking about the sunset period for TV — at least as we've known it.

The dominant role once played by a few TV networks is being transferred to new forms of media in a 500-channel universe and rapidly expanding news distribution systems. The growth of hate on the Net is a concern for all justice-seeking people, yet few governments or regulatory agencies have been willing or able to prevent the hateful messages from reaching larger numbers of home computers.

Governments seem unable to stem the growth of certain dangerous forms of pornography on the net. Lies, untruths and gross opinions abound — and **it can be like walking into Toronto's SkyDome with 30,000 people talking at once.**

Yet the Internet is influencing mainstream journalism in all new ways. Shortly after the TWA 747 explosion off Manhattan Island, dozens of conspiracy theories emerged about bombings by terrorists, friendly missile fire gone astray and others. Respected journalists got sucked into a vortex of wild ideas and their reputations suffered as a result.

One result of the Internet is to break down the barriers between insiders and outsiders. With access to so much information and so much power to inform, the lines become blurred. Yet, more than ever, when we say something, it leaves a trail that invites others to take up the message.

More and more, we're communicating under a microscope. And maybe there's something to be said for a society that is not afraid to live more openly.

Topics that mainstream media avoided — like the sexual escapades of certain public figures, are now widely reported and discussed on the Net. Are the reports true? Is the Net the communications panacea of the future, or do you prefer the editorial board of your daily newspaper to make the tough decisions about what's newsworthy and what's not? And who do journalists work for, anyway?

The experts in change management agree on one thing — that the language of change is often blunt. We can view the news media in many ways — as evil forces of capitalism involved in an international conspiracy, or as conscientious leaders of social change, but chances are, we can't change the news media. We can, however, change our message management in order to take advantage of how they operate. We call it *MediaSpeak.*

In a review of the above CBC documentary in the *Toronto Star,* esteemed media reporter Antonia Zerbisias writes:

> "Still, *Dawn of the Eye* does provide ample evidence that, not only did the camera change the way we saw the world, it changes the world we saw. Sometimes for the better — as it most certainly did in holding back the Chinese government from slaughtering every student in

Tiananmen Square — and sometimes for the worse…

The camera altered the course of history when, in 1960, US Vice President Richard Nixon sweated through a televised debate with the more telegenic Democratic contender, John F. Kennedy…

In Canada, a weekly television show helped create a star out of René Lévesque, who would go on to lead the Parti Québecois. And former Prime Minister Mulroney took media wrangling lessons from President Ronald Reagan, whose advisor, Michael Deaver, boasts in *Dawn of the Eye* that he suckered the Washington press corps (and the nation) for years with his daily staging of camera-ready visuals that cast his boss in the best light.

'I basically became the producer,' says Deaver. 'I tried to produce every day something that was so irresistible to the networks that they couldn't turn it down. I did their work for them.' "

And Reagan was called the great communicator! Ronald Reagan was the product of high-priced packaging. So was former Canadian Prime Minister Brian Mulroney. So are the banks.

What was Clinton's strategy?

People's opinion of public affairs and events is shaped, in large part, by the news media.

In order to deal with the news media, we need to manage our message — the strategy, the choice of words and the delivery of those words.

In my humble opinion, I believe that Clinton's early strategy was to stick to a script based on at least three areas of focus:

"It's the economy, stupid!"

As long as the economy was strong, voters didn't seem to care much about the private lives of their elected officials. Besides, many politicians have been philanderers and the public continues to worship former President John Kennedy.

Jimmy Carter, on the other hand, only committed adultery in his heart, and many people then considered him to be a wimp. Apparently, voters don't even much care if their politicians always tell the truth, since few believed candidate Clinton when he said he tried smoking marijuana, but he didn't inhale.

"The Hillary factor"

If his wife is not publicly upset about his affairs, why should anyone else care? Even though Hillary denied acting like Tammy Wynette and "standing by her man," that's exactly what she did. She, too, had a job to protect, as well as a home at 1600 Pennsylvania Avenue with a few perks.

"The Bimbo factor"

What's interesting here is that Clinton's support among women was much higher than among "angry white male" voters. Maybe there's a jealousy factor at work here. It's been said that a promiscuous person is "anyone who gets more sex than you do." It's a sad, unfortunate reflection on our gender-bound society when a man is sometimes still admired for his sexual escapades while a woman who performs oral sex is considered loose.

What was Clinton's method of delivery?

Could you go on international TV, direct-to-camera, live, before an estimated audience of 60 million people and admit that you hadn't told the truth about your sex life? Clinton admitted that he hadn't told the "whole" truth soon enough, confirming a basic principal of managing your message in a crisis: tell the truth quickly, tell it well, and tell it again. This becomes the *script* of your message.

Clinton's messages have always been highly scripted. After all, he had access to hundreds of advisors, spin doctors and assorted hangers-on. There have even been suggestions that Hillary was involved in preparing some of what he said. In any case, he stuck to the script like glue.

Here we have a country where a former professional actor (Reagan) became a politician.

Clinton is a professional politician who became an actor.

Clinton, who can and often does talk all day, is the master of the sound-bite. Clinton's acting lessons prepared him to deliver his lines.

While Ronald Reagan's staff always denied that he had his hair colored that unique shade of purple, Clinton's staff let slip that he had his hair shaded grey to appeal to older voters.

Ronald Reagan wore brown suits with the pockets sewn shut to keep him from sticking his hands in them. Clinton's suits were said to be test-marketed on focus groups before he wore them in public.

Waving at trees

According to The Globe and Mail, Brian Mulroney and his wife Mila, learned how to wave at trees. Their teachers included the Reagans.

It works like this. When getting off an airplane or descending a platform, all the cameras are aimed right at you.

The cameras can't tell that you're confronted by angry and hostile protesters or that the crowd is very small, the cameras only show you waving. So, wave to both sides, even if you wave at trees. And keep smiling, even if the crowd isn't. Produce that photo opportunity for the evening news. Look like a winner.

The Mulroneys also mastered the trick of pointing at members of the audience as though they were their oldest friends, while at the same time touching their spouse's elbow to get them to join in the waving.

There's no business like show business and there's a sucker born every minute, but you can only fool all of the people some of the time.

As Brian Mulroney often liked to say, "There's no whore like an old whore." He would know.

Chapter Two

Packaging the message

It's about packaging

The news business today is largely about the packaging of info-tainment.

Shouldn't you apply some of those same skills to packaging yourself, your organization and your message? Think of a media interview message or any other communications encounter as an infomercial, in which you have full control over the production at all levels. What would you say about your topic or issue:

- if you only had one word?
- if you only had three words?
- if you only had 12 words? (that's the equivalent of about 5.2 seconds of air-time)
- in a 30-second message? (where you'll have about 65 words)
- in a 60-second message? (you'll have about 125 words)
- in a seven-minute one?
- in 30 minutes?
- in 60 minutes?

The first skill for any spin doctor is to learn how to count — in time and space. Then you consider the format — how much editing will take place to the material. How much slicing and dicing is going to take place in the editing suite?

Have you ever been in an editing suite? Today's advanced computer technologies allow for fast production of print and electronic packages, and if you're responsible for managing your message, you'll have to learn some basic skills to use when being interviewed.

If you're producing your own infomercial, you should have a professionally-written script, lighting and camera angle instructions, and access to graphics and other enhancements. You'll need to learn how these work in the newsroom, too.

A typical TV news story on the six o'clock newscast will run for 60 seconds. It will likely contain one or two **"talking-head sound-bites, actualities or voice clips"** that will average about 5.2 seconds each — that's about 12 words. The reporter will be seeking balance and will attempt to illustrate at least two sides to the issue. A number of editors will be involved.

In order to write a headline or the lead paragraph in a front-page newspaper story, the reporter only needs one word — generally a glaring word or the denial of a glaring word. Shouldn't that word be one of your choosing? What one word describes your concern, your role, your mission, your plan or your team? What words describe the reputation you would like to have? Welcome to the world of spin.

Preparing the package

If we look at spin as an opportunity to package and thus influence the message, we become the directors, producers and performers in our own infomercial that we package for the news media.

The better we package our *MediaSpeak* message to meet media standards and style, the less likely the news media will be to change it, and the more control we'll exercise over the resulting story.

Interview preparation can often include rehearsals in a studio or with a trainer who will impersonate a reporter. My students call me "the reporter from hell." In preparing for a candidate's debate, actors or staff will assume roles of each participant and there are sure to be some planned surprises.

You'll have to be prepared for an ambush by your opponents or by a reporter who assumes the role of your opponents. During the first few seconds of your reaction, your reputation hangs in the balance — no matter what else you say later. Your 5.2 seconds of fame have arrived.

If we can package our *MediaSpeak* message in an appealing, interesting and concise manner, we've met the media's needs. Our success, then, comes when the final media message is delivered as we intend it to be.

But if we package our message poorly — if it's too long or too technical — if it's not written in plain language — then the reporter or editor must re-package it to make it newsworthy. **The result can be misquotes, untruths, distortions, unplanned controversy**. Others may succeed when we fail, gaining the advantage of their superior spin. Packaging starts with the old theory — garbage in, garbage out.

If you're being interviewed for a technical or professional publication, you'll be able to increase the technical level of your language. If you're being interviewed for the mainstream TV audience, craft your message for a grade six comprehension level. Bafflegab and jargon can damage your delivery.

Once your script for an infomercial is complete, you'll have prepared most of the answers needed for a reporter's interview on that topic. A 30-minute infomercial will provide you with more than enough material for a 30-minute interview. This book will teach you some of the basic skills needed for that preparation.

We'll examine how the news media packages information so that your packaging meets their needs, and yours. If they want 5.2 seconds, give it to them, in *MediaSpeak*.

If it bleeds, it leads

Reporter Zerbisias continues in her review of *Dawn of the Eye:*

> "During the Dirty Thirties, when bread lines snaked through city streets and Nazi Germany was terrorizing its Jewish citizens, the newsreels concentrated on fires, crashes, cute kids and bathing beauties. No wonder that a wag then observed that newsreels were nothing more than a 'series of catastrophes ended by a fashion show.'
>
> 'That is pretty much a good definition of any local American newscast these days,' says (Mark) Starowicz (of the CBC.)"

One wonders why Mr. Starowicz confines his remarks to **American** television news when Canadian news programs aren't much different. In some Canadian TV newsrooms, the motto appears to be, "If it bleeds, it leads," referring to the trend of leading the newscast with the most recent car accident or flying truck wheel story — often because the station happens to have video footage.

If a tree falls in the forest and there's no tape recording…? But what if a tree falls during Ice Storm '98 in Eastern Canada, and a TV crew is taping it? The graphic TV footage showed us all how the laws of nature, physics and probability can come together. Reporters stood in the middle of Montreal streets as trees and power lines fell behind them, as if on cue.

What makes you newsworthy?

Clout. Credibility. Influence. Who has it and who doesn't? Who decides who has it and who doesn't? Is life fair?

Our actual level of importance, our official job title, our reputation, our level of responsibility can change in the instant of a media encounter.

Some experts use the term "**icon**" to describe the symbolic role attached to any individual, who for however brief a moment, represents to the public an issue, situation or topic. Whether you're a household face or an unknown, for 5.2 seconds you're a public representative, to be interpreted or dissected by journalists you may never have encountered before and may never encounter again.

For one, brief moment, you may represent your entire hospital and hospitals everywhere. For 5.2 seconds or less, you're the most important accident investigator in town. In under 15 words your career is on the line. One wrong word and you're toast. Fired. Canned. Put on special assignment. Forced to work in Ottawa for 18 months. (It happened to me.)

What makes you newsworthy is the spin you bring to the story, what you can contribute, your information or your opinion — along with how that message is delivered.

You may be battling city hall to stop a street expansion in your neighborhood, or you may be a member of the transportation project team handling street expansion. Reporters will want to talk to both of you because each of you adds a different dimension to the issue and the news story.

You might be a healthcare worker involved in drug education programs or a shopping center manager involved in the issue of teens hanging around your mall dealing drugs. Reporters will want to know at least two sides to this issue and talk to teens, parents and other "experts."

You may be a police officer explaining an emergency situation to reporters at the scene of a major traffic accident. You may be a witness to the accident being interviewed by reporters. The witness, as private citizen, has very little reputation to lose in the encounter, yet one wrong word from the police officer can result in lawsuits.

You might be a public affairs practitioner for a community agency like the United Way. Or a union spokesperson concerned over the issue of work for welfare among United Way Agencies and you're calling for a boycott of the United Way.

Or, **you might be someone whose family is aided by the United Way**, as you become, for a brief instant, a public example or symbol of United Way services at work. If people respond to your situation the wrong way, donations might fall.

You might work for a manufacturer that has been hit with a product-tampering crisis — or you may represent a consumer group calling for better packaging and container instructions on that company's products.

You might run a restaurant where disabled people are complaining about access to the washrooms—or you might be a disabled person writing a letter to the editor about this situation. And that letter to the editor might prompt a reporter to contact you for an interview—positioning you as an instant expert and a voice for the disabled, or differently-abled.

You may be the executive director of a national organization, like the Royal Canadian Legion, talking about the issue of wearing turbans in Legion halls. Or you may be the local club manager instructed to outline your own club's position. Or you might be a Sikh veteran denied entrance to a Legion facility. Each of you brings spin to the story. Each of you has a different level of reputation management responsibilities.

You might be a politician running for public office, or a staff communications specialist responsible for keeping your political boss out of trouble by always saying the "right thing." **By being on the "right side" of the issue. By never letting anyone else "out-concern" you.**

You may be a cabinet minister, able to demand three different media messages, from three different staff members. You can choose what you feel is best or blend and combine advice from the experts, keeping in mind that the final choice is yours. Or you might be one of those staff members churning out briefing notes so your boss will keep her job.

You may be working to manage an emergency situation where you have a hundred operational responsibilities and dealing with a roomful of reporters is not your favorite duty. Yet you know that if you don't communicate the seriousness of the situation, people will not send emergency supplies or cash. A tip of the hat here to the Oklahoma City fire department for their compassionate, articulate and professional spokespersons after that city's terrorist bombing.

Your emotional quotient

There is a formula for who makes the best spokesperson. I'm sure it's the same formula for who makes the best family doctor, the best pilot, the best police officer, the best banker and the best President or Prime Minister.

This formula combines the following attributes: caring, speaking out and the ability to proceed in a co-ordinated manner. Or, as described in recent news reports, your emotional quotient. **Your EQ is a measurement of empathy, assertiveness and crisis management skills.**

Your EQ can be illustrated in sound bites. Here are some examples:

> **"Of course we're concerned about the environment, that's why..."**
>
> **"People are upset about this new policy, that's why we're..."**

Case Study: Caring about squeegee kids

In my rather humble and completely unsolicited opinion, former Toronto Police Chief David Boothby made a strategic error in managing his message about the popular media stories on squeegee kids in Toronto in the 1990s.

Boothby stated that "poverty was not the issue." This created lots of controversy.

His error was in saying what the issue was "not." Of course poverty was a factor in the issue. He couldn't make that go away by saying it was not important. Instead he should have focused on police issues that were inarguable or were least arguable, like public safety.

He immediately enraged poverty activists. Reporters ran out to do stories on the horrid living conditions of a few of the kids. The police chief came across as uncaring and unaware, and he could have avoided that easily with an inarguable premise.

Here's what he should have said:

"Yes, some of the young people are facing poverty, that's why it's important for them to work with social agencies who can help them. ***Our main concern at the police department is public safety.*** *Let me give you an example of an incident that occurred on..."*

Write a letter to the editor — today!

If you've never had any dealings with the mainstream news media, there's an easy place to start. Write a letter to the editor of the largest daily newspaper in your community.

Pick a topic that interests you, like transportation, education or one of today's front-page stories. Assume everyone you know, including your boss, will read the letter and be impressed with your brilliant way with words.

Letters to the editor are among the best-read parts of a newspaper — an early form of chat-line. They are free. But if you stray from a formula, they're less likely to print them. They are most likely to publish your letter if it meets their criteria.

Here's a few tips on getting a letter to the editor published.

☞ *The final recipients of your message will be other newspaper readers. Tone your letter accordingly. Do not use attack language.*

☞ *Write about a specific item headlined in the newspaper, example: Re: "Women win court challenge on pay equity laws" Sept. 14, 2005*

☞ *Find creative links to your issue.*

☞ *Don't get mean, petty or angry at the editor. Make it light, clever and interesting.*

☞ *Do not write generic letters to all media.*

☞ *Keep your letter under 150 words. The trend is toward ever shorter letters, so comply.*

☞ *Keep words, sentences and paragraphs short, just like a newspaper article does. Write in MediaSpeak.*

☞ *Send same-day as item appears or ASAP, by facsimile, e-mail or courier, but do not pester anyone.*

How big is your budget?

You've all heard the joke about the lawyer who was asked how much he charged. In response, he asked, "How much have you got?" That's the first question PR people often have to ask.

In mounting a media campaign, your budget may consist of the cost of e-mail, sending a facsimile transmission or posting a snail-mail letter. You can make a difference with one letter to the editor, which can cost less than a dollar to send.

If your communications budget allows, there are wider opportunities for influencing news coverage.

Case Study: Caring for cleanliness — Sunlight Soap

Recently, a company making dish-washing soap produced a video news release featuring the cleaning of the roof of Toronto's SkyDome with their product. The company issued the video news release through a service that reached 7 million viewers on the evening newscasts on over 40 TV stations world-wide. The TV stations ran the news release as a news item, at no charge, of course.

There are several companies that will issue your news releases or news packages for a fee. You can choose from a wide range of services targeting key media in a number of sectors, including video news releases. As for costs, the SkyDome's the limit — simultaneous video news-conferences around the globe, free helicopter rides for journalists, whatever your budget will allow.

The production costs for the "soap spot" were reported to be about $10,000. The great photos of cleaners toiling on the SkyDome roof made the story attractive to the newsrooms, and it didn't cost them a cent to send a news crew to the CN Tower to videotape the event.

The question remains — **are we blurring the line between news and infomercials? Yes**, it's happening every day. Journalists continue to debate the relative merits of using these services, yet the

Toronto Police Services operates a video production unit to assist the news media.

Toronto also has TV cameras mounted on highways to provide information on traffic conditions. Highway cameras have been known to capture some intimate moments on the highway shoulders.

In the 1992 US election, Bill Clinton's media staff changed the way the news media covered politics. Armed with a software package that could reach over 300 newsrooms simultaneously, the Clinton "war room" was able to respond to opposition attacks with prepared sound bites in under 30 minutes. The term for this method is "negative questioning response team."

There are budget and strategic decisions made in newsrooms every day. Does the newspaper run the stock photo of the politician on page twelve, or do they send out a photographer to wait for him to pick his nose at the dinner table and run the resulting shot on the front page? Do they cover your new product or carry a wire service story about a competitor in another country — because it's cheaper to carry the wire service story?

Those with the budget clout have a distinct advantage in this business, but like David and Goliath, **well-written words can bring down the biggest opponent.**

News is what's new — even when it's old

Incidents, accidents and disasters are constantly occurring and many of them bring to light larger issues that may capture the attention of the media — especially if people are talking about the issue and calling for changes in policies, laws or practices. What is news? Whatever editors decide is news, that's the oldest definition. How much attention an issue receives will depend on what else is happening that day and how much public interest there's likely to be.

What's the difference between a situation and a crisis? A situation is "under control" — or at least it appears to be. That appearance often depends on spin — what we call message management. Your reputation depends not only on what you do and say, but what is being reported and said about what you do.

A few years ago, who would have predicted that truck tires would fly over Ontario highways, killing five people and thrusting the families of some victims into the media spotlight? Some of these family members have proven to be articulate, compassionate and readily available spokespersons on whom reporters can call at a moment's notice for a "quotable quote" on issues of truck safety.

Yet, before the truck-tire tragedies, these individuals could never have imagined the new roles the media bestowed on them. Are you the voice of an issue? Are you prepared to become an icon? Welcome to the public chat-line that is the news media.

No matter what issue, topic or concern propels you to deal with the media — or propels the media to want to deal with you — you'll want to deliver the best possible message in a concise, interesting and direct manner.

Be prepared, professional and pro-active.

A reputation for speaking well

You may be a natural talker or very nervous in public speaking situations.

You may be able to handle your responses spontaneously (which is what the media are looking for) or you may need to stick to a carefully plotted strategy using a written script or notes. There are three ways to deliver your message — reading from a text, speaking extemporaneously from notes or ad-libbing. Ad-libbing is most likely to create problems, unless you have a wealth of experience and subject knowledge.

You may want to give the impression that you are purposely sticking to a prepared line, or you may want your stock responses to sound natural and spontaneous — even when they're not. When a spokesperson says, "It is the company's position that...," the spokesperson may be distancing him/herself from the issue in a strategic way.

There's a very fine line between issuing worn-out clichés and coming up with an earth-shattering sound-bite at the right moment. What's creative one day can be shop-worn the next. When US President Bill Clinton told opponent Bob Dole, "that dog won't hunt," he grabbed at our emotions with a metaphor. The same when Walter Mondale borrowed a line from Wendy's Hamburgers and asked Gary Hart, "Where's the beef?" How often did we hear Clinton's slogan "bridge to a new century," yet can you remember any of Bob Dole's one-liners?

There's a big difference between saying very little and saying very little in an interesting and informative way. It's one thing to avoid making a commitment, it's another thing to pass up a chance to illustrate your concern for the public and at the same time display your professionalism.

It can be very frustrating to watch a politician avoid answering a question, but the experts suggest that a lot of our reaction to the interview depends on whether we like the politician in the first place.

If we like her, we will feel as though she's positively assertive and capable. If we dislike her, we'll consider her evasive and manipulative. If we have not decided about the person, we can be influenced in either direction depending on the quality of the delivery. Some politicians do a lot better job than others in "not answering" the question.

Who's your boss?

This is often the second question a PR person asks the client.

Your media message might be one you develop all on your own, for which you take full responsibility. If you're your own boss, you're authorized to use all the rope you need to hang yourself.

Don't you still want to look professional and public-spirited? Consider the alternative image — looking amateurish and self-centered. Choose one.

Or, your job might require that you speak to the news media on behalf of your employer. Is your job on the line? As US Attorney General Janet Reno once said during the Waco, Texas crisis, "My job is always on the line." Reno was a top-notch spokesperson who was careful to stick to a planned message. She didn't get where she is by ad-libbing.

As any speech-writer knows, it's almost impossible to write good copy by a committee. When several people have to approve a message, each of them feels obligated to make changes. There are conflicting views about what to say and how to say it. Only when you learn to brainstorm creatively can your teams manage messages effectively.

Yet even if you have a speech-writer, the responsibility for the words is still entirely yours. Ted Sorensen was speech-writer for US President John Kennedy. In an article in *The Toastmaster,* June, 1998, Sorensen said, "John F. Kennedy was the author of all his speeches, because the decisions as to what subject matter, what policy, what decisions to convey, were his." According to Sorensen, "group authorship is a recipe for poor speeches." He could have added media messages, too.

Here are some questions to ask yourself very early in the process:

What do I hope to achieve as a result of my media encounter — personally and corporately?

What if I only get one opportunity to say something to the world in one or 12 words?

What will I say if they ask me about this topic?

What should I be stressing and what should I avoid?

A phrase like, "We're sorry for the inconvenience," may be as important in many media strategies as it is in customer service language. Or maybe the only answer for the present situation is "No decision has been made on… What we're doing is (this, this and this)."

When the same old questions occur again and again, develop "stock responses" to use again and again. No matter how often you're asked, deliver the answers freshly and with the appearance of spontaneity.

Remember, spokespersons should position themselves so that they illustrate their responsibility to the public as well as to their employer, which places them in a unique position in an organization.

When you remember that the public is a client, you'll remember to design messages for that client. It's a common strategy to **position all your statements with a major proportion of public-spirited professionalism.** (Note the alliteration.)

Some spokespersons are provided with an approved message from which they cannot vary in any way — and sometimes it's essential to read from a prepared statement in highly controversial legal situations. It can be very uncomfortable reading someone else's words as if they're your own, unless they're well-written. A committee can't always achieve that.

You may not need to re-invent the wheel to develop media messages, **but you should say what you intend to say, no matter what you're asked.**

Remember: It's seldom easy to keep it simple.

Chapter Three

The message will set you free

Introduction to *MediaSpeak*

MediaSpeak is designed to equip you to handle the toughest questions. You start by answering the easy questions.

What's your main concern(s)?
What's happening, or has happened, or will happen?
Can you describe the situation?
What are your plans, priorities or reasons?
What are the main factors in this issue?
What will the impact be?
What needs doing?
Why?
How?
More specifically?
How much?
By whom?
When?

Drafting *MediaSpeak,* or media lines, requires you to anticipate the very worst possible questions and have answers ready — or deflect or defer the questions, while controlling or re-directing the interview back to your specialty. When you have answers for the easy questions, you use them to answer the tough questions, like:

Have you ignored...?
Why are you dragging your feet?
Is this politically-motivated?
Are you price-gouging?
Is there a secret agenda or cover-up?

We'll introduce you to pillars, supports and sparklers — our *MediaSpeak* tools for writing media messages that match what the media are looking for. Once you have the answers for the easy questions, you'll learn to use them to answer the tough questions. Spontaneously.

Is the medium really the message?

Consider yourself the medium in the interview — after all, you're delivering the message. Consider the reporter as a vessel, through which your message flows to the final recipients.

Andy Warhol once said that everyone would be famous for 15 minutes. **I believe that it's more likely to be 5.2 seconds** — about 10 to 12 words. I also believe that you should decide what you want to say in those 5.2 seconds, rather than be trapped by a reporter who puts words in your mouth. Or a reporter who presents questions in such a way that you lose control of your agenda, or you use words that inflame an issue.

The question you have to ask yourself is: "Do I want to go where the reporter is trying to take me?"

Would you want to be quoted saying:

"We're not ignoring anyone."
"There's no political motivation here."
"There's no price-gouging."
"There's no hidden agenda."
"This is not a poverty issue."

All of the above answers are presumably true. But do you want to be quoted using those glaring, inflammatory words? Can a reporter trap you into using those words? It's the most common error among my clients and it's one of the most important teaching methods I employ in preparing students for the reporter from hell.

Each of the above statements is the denial of a **negative**, resulting from the wording of a question.

You may want to use glaring and inflammatory language to attack others. But what if you're being attacked?

Your goal may be to inflame the issue — or to control it. How you deal with the issue depends on your words, your delivery and your strategy.

MediaSpeak is designed to provide the spokesperson with a package of core statements, background and context that match the format of the interview or the medium. If you're being interviewed for tonight's six-o'clock local TV news, the reporter will likely be using between two and 10 seconds of your entire interview in a "talking head quote."

You'll want to be equipped with several key statements of about 5 to 25 words. These are designed to become your "quotes" and you'll confine your message in order to influence what quotes are used.

"Welcome to the evening news. I'm Ian Taylor."
(I've always wanted to say that)

The six-o'clock news story is the benchmark media encounter to which I train most of my students. (In our two-day Manage Your Message workshops, we prepare students for this particular interview format, along with a live-to-air interview, also for TV. Then we adapt that delivery to other formats, such as newspaper, radio, in-person and telephone interviews.)

The TV reporter probably knows much of the story before arriving for the interview.

The reporter is looking for one or two sound-bites from you. And the same from someone else involved in the issue, to balance the story.

Too often, what the reporter is looking for here is controversy. Will you contribute to it or prevent it?

The reporter wants stand-alone answers, that don't need the question repeated in order to make sense.

The reporter wants clarity, conciseness, interesting material that is informative and entertaining.

The reporter wants emotion from you — pompousness, defensiveness, guilt, embarrassment, anger on one side. Or, empathy, concern, understanding, knowledge, caring on the other side.

The reporter will work with the photographer to gather visuals, including your talking head shot. They want B-roll (background) and shots of you over the reporter's shoulder (talking heads.)

Avoid interviews at your desk. You create an artificial barrier between yourself and the viewer.

Stand up, preferably outside with relevant background that will position you as a hands-on spokesperson in touch with the issue.

Wear clothes that have no pattern, except the tie or scarf.

Direct your eyes at a spot between the reporter's eyes and the camera lens and focus on that — the viewer won't know the difference.

Look at the ground during the question, count to two at the end of the question, deliver your message.

Where does TV get its ideas?

Many radio and TV news directors determine their story priorities based on what's contained in the morning newspapers. This is a form of **pack-journalism** — where everyone is following the same few stories of the day. Many news outlets carry stories because others are doing so, or because news agencies are packaging the top ten stories of the day. Newspapers love exclusives, and that's where you should market your story if you want to make it later on TV.

I once took the tourist's tour of the world's largest TV newsroom — CNN in Atlanta. It was no surprise that almost every desk in the studio had on it a copy of that day's local newspaper, *The Atlanta Constitution.* There's an old saying: "Where does TV get its news? From the newspapers." And there's a lot of truth to it.

TV news magazine programs like *60 Minutes* can also generate print coverage Monday mornings when they break a large story Sunday evening. The tabloid TV shows and the endless stream of TV talk-shows generate new material but they get most of their ideas from print, and increasingly now, from the Net chat-lines.

It's no coincidence that a cover story in *Time, Newsweek* or even *Sports Illustrated* results in copy-cat coverage in other media. The breaking story contains all the research and a lot of background and other reporters just do a follow-up, find a local angle or summarize the main story.

How Oprah selects her guests

TV and radio talk shows hire producers to screen possible guests.

The producers carefully interview the guest by telephone to see how they answer questions. What are they looking for? Spontaneity. People who like to talk. People who speak conversationally. People who sound interesting. People who employ techniques like pillars, supports and sparklers, which form sound-bites.

If you fail the test, you won't get on. If you pass, you may have a new career.

Do you know how many books Oprah can sell in a one minute review? Neither do I, but I'd like to find out.

Copycat journalism

As newsrooms become smaller, the copycat trend is increasing, and disturbing, because it means that so many worthwhile stories are not being told. Or that the same stories are being told the same way by many media all day long.

Sometimes reporters miss important parts of the issue because certain agencies are stage-managing their participation so smoothly and effectively. Some spin doctors are doing a lot better job than others at managing their message.

Remember, you can't likely change the way the media operate, but you can change the way you operate with the media. As the old sailor's expression says, "**If you can't change the wind, change the sail.**" If you haven't been successful in getting media coverage, change the packaging and/or change what you're doing. It's a competitive business to attract media attention.

Newsrooms often don't have the staff to dig out new stories or attend every news conference or follow-up on every news release they receive—no matter how newsworthy it may be. At one time, reporters often produced "enterprise" stories — where the reporter dug out the material in many different ways and presented the finished report to an editor.

Today, with fewer reporters working in newsrooms, the editors are often making almost all the decisions, including how and when and where the story will run — or if it runs at all. That means more and more groups and individuals are competing for scarce media attention and in order to succeed, your message must be packaged in a way that meets the media's requirements. It must be concise, interesting and direct.

A story takes on a life of its own

The breaking story carried in this morning's daily newspaper can start to take on a life of its own as various people and organizations react to it. And people will react — with telephone calls, facsimile news releases or Internet messages. Today's technological advances can permit organizations to send out hundreds of media messages simultaneously to hundreds of news agencies.

Editors love a five-day story. An incident takes place, related events occur and more organizations become involved. The viewer/reader expects the morning's news to carry the latest chapter for up to five days, depending on the story. Each day, the spin can change. If you're involved, you'll have to work either to maintain or amend your core messages.

Questions may be asked in the legislature or at city council as a result of the coverage. Groups will suddenly see new opportunities to advance their issues.

Public interest groups will begin to become involved or see opportunities to further their own agenda. Groups representing professional organizations or business interests, unions, neighborhood associations, aspiring politicians, church groups, government agencies and "concerned citizens" may all have a role to play in certain issues.

The evening newscast can then provide the public with the latest update on what they read in the morning newspaper, but with added touches as new participants enter the arena. And tomorrow's morning newspaper will have a new chapter to kick-start the cycle.

Many of these issue-participants will be well-trained in handling the media. They'll be interview-savvy and aware of what hot buttons to push, or what key words or phrases to use to capture the media's attention, or to deflect that attention elsewhere. And it may not matter how credible these organizations are. It may depend more on the quality, suitability and newsworthiness of their *MediaSpeak* message at the moment.

In today's news business, it's a matter of **time, chance** and **space.**

The effect of the global village

Newsrooms monitor each other constantly for story ideas and trends. An editor at one news outlet will check on the competition to see what they consider newsworthy. If the other news outlet has found an interesting spokesperson to interview about a topical subject, that spokesperson may get calls from other reporters.

Suddenly that spokesperson is in demand, sometimes traveling from obscurity to the front page because journalists consider what they have to say is worth repeating. The spokesperson becomes, for however brief a moment, an icon.

A major story in Chicago about shoplifting by a prominent senior citizen, for example, might prompt reporters in your community to seek out a local angle.

Suddenly you might become that icon — as an articulate and relevant voice for seniors or for the retail or security industry. The story has lots of possible angles: Age, race, poverty, living conditions, costs of prescription drugs, loneliness, government services, healthcare, retailers' policies, security equipment and video monitoring, police policies and more.

That's why it's essential for public affairs practitioners and potential spokespersons to monitor daily news coverage — to be aware of what we call "**the public opinion environment**."

This is the collection of all the coverage (news, opinion and analysis as well as how much prominence the item receives) and its anticipated impact on public opinion and on your reputation. I call it your reputation vulnerabilities.

Will this information change or reinforce public opinion? Will information result in changing policies, actions, decisions or laws? Should it?

When the Brian Mulroney government hinted at changes to the country's pension plan, it took one senior citizen only a few seconds to turn the issue around. TV cameras captured her telling the Prime Minister that if he touched pensions, it would be "Good-bye, Charlie Brown." Talk about the power of two seconds! One senior citizen, for that brief moment, symbolized Canada's elderly and everyone's mother. God bless her.

Professional spokespersons may spend a good portion of their day reading newspapers and magazines and tuning into broadcasts of radio and TV. They're monitoring public opinion. They're learning new messages. They're seeking opportunities to apply their spin to the story, or stay out of it. They're hopefully learning what not to say, too. Or where not to grant an interview. Or what to ask the reporter before agreeing to an interview.

Some questions to ask a reporter before agreeing to an interview

What is your name, news outlet and phone number?

What is the story about?

Could you be a bit more specific?

What do you want to know?

What will be the interview's format — or how much editing might take place?

When do you need to talk to someone?

How long do you need to talk to someone?

Are there visual needs for TV or newspaper?

When and where will we conduct the interview?

Would you object to my tape-recording the interview?

What not to ask a reporter

Can I see the story before it appears?
Only if your name is Rupert Murdoch, Ted Turner or Conrad Black and you own the news outlet.

Who else are you talking to?
Instead, ask if there's anyone you can refer the reporter to, and have someone in mind that's ready to roll.

Can you send me a copy of the story?
Get it from a media monitoring company or make other arrangements. Reporters usually forget anyway.

Case Study: Caring about safety — Flying truck wheels in Ontario

You may be a member of a family whose daughter has been killed by a flying truck tire — or the president of a large trucking company.

In either case, you may be of equal importance in the eyes of a reporter seeking balance and fairness in preparing a story. Each of you must have a strategy, a goal to achieve in your interactions with the media.

Your first goal may be to keep your job, or your goal may be to change the laws, or to change safety policies, or to outline the thoroughness and professionalism of your company's existing safety programs which are leading-edge.

But if your trucking company has no safety program, no commitment to safety and a record of violations, you deserve all the negative attention you get.

Your strategy, therefore, should be to admit that there have been problems in the past and that's why you're improving programs. No truthful media line will save your sorry butt when you deserve to be in jail. And anyone issuing a media line that is not truthful is asking for trouble. Once you've cleaned up your act, you'll have lots to say that will reflect on your organization's public-spirited professionalism.

Your author is the son of the owner of a small Manitoba trucking company which operated from 1907 until 1980. I've trained over 300 truck safety experts and have the highest regard for most of them.

The issue of transportation safety will not go away, nor should it. It only takes one airplane, snowmobile or car accident to place the issue back on the front pages. There are dozens of ways for you and the media to approach this issue which show the need for a planned message, delivery and strategy.

Assume for a moment that you are the designer of a new instrument or device which instantly shows the amount of wear on a truck's brakes. A small item in this morning's newspaper (from your company's news release perhaps) catches the eye of a TV assignment editor.

Even though it's a fairly busy news day, the issue of truck safety affects large portions of the viewing audience. The assignment editor instructs a reporter to do a story for tonight's news.

The television reporter's task may be to develop a news story that will average 60 seconds in length or about 125 words, for the 6 o'clock news. The reporter will want to acquire one or two quotes from you, in addition to what they've received already in your news kit.

These quotes should be from your planned media line that you prepare once you know what specifically the reporter is seeking and once you've prepared yourself for the widest range of questions.

The reporter will also attempt to interview someone with a contrary view, since many stories consist of conflict or controversy. Reporters seek at least two sides to every story.

The contrary view may be from an independent trucker who says your device is too expensive and not necessary if truckers carry out a visual check. Or the reporter might seek out a university engineering professor to examine and evaluate (however briefly) your invention. Or seek out a competitor whose goal is to sell their product by criticizing yours.

Even though you may think you can control all the coverage, you can't. You can, however, control your part of the message with professionalism and in a public-spirited way.

If all else fails, the reporter can go out in front of your office and ask "the person on the street." No one may ever know the exact wording of the questions asked. There's no science here, but it appears to be public opinion. It's not. One news editor told me this news method is called the "asshole magnet." Apologies all 'round.

Media requirements are based on space and time

Most of us speak at about 125 to 150 words per minute or about two to three words per second — which varies based on rapidity of speaking, planned emphasis and pauses purposely placed in your delivery.

A 10-second clip runs about 20 to 25 words. A 15-second clip — around 30 to 40 words. Most news stories will be subject to editing by the reporter, and the reporter will use maybe one or two of these quotes or voice clips in the entire story, even after a one-hour interview. **Won't you want to provide the reporter with tantalizing messages that are likely to become quotes?**

News stories are prepared to meet strict requirements for time in broadcast media, and space in print media. **It's therefore essential for any writer/spokesperson to count your words and time your messages to meet the media's requirements**. Learn to count.

Most TV news stories on the 6 o'clock news run no more than 90 seconds — a total of about 200 words the reporter must compose to tell the story along with visual background and interview sound-bites, quotable quotes and talking-heads.

The final product bears very little resemblance to the way it was originally assembled, except for the live-to-air broadcast. The interview will be edited. Diced, sliced and re-packaged into seconds or words. Often, the question gets edited out of the final product. **It's not the question that will get you into trouble, it's your answer.**

Most news stories in a daily newspaper run about 150 to 400 words. That's about four to 10 column-inches. Feature articles can run from 1,000 words and up. Some reporters gather their information in one-hour interviews, others can grab three quotes in under two minutes. Your goal is to give them at least seven quotes in your first two minutes.

A newspaper columnist is required to file a specific number of words per column — often about 600 to 900, depending on how much space is available. If the columnist only writes 500 words or files a 1500-word piece, there may be a job opening for someone who can do what they're told. The editor is saving a space for a certain number of words. Advertising space has been sold two days before and it has priority on a newspaper page. All media activity is measured in space and time to fit between the advertising.

Your media line may only consist of 250, 600 or 900 words. And if it is as well-written as the best news story, if it is concise, direct and interesting, if it is delivered well — then some or much of it may become news and that's your goal.

How long is my media line? How long is the interview?

A media line should equip you with enough material to match the length of the interview to which you agree to participate, just as a 15-minute speech is designed for 15 minutes of platform time. The more high-profile the situation, where you may be getting calls from dozens of reporters about the story of the day, the more important it is to be **consistent and concise and repetitive**. We'll teach you to prepare pillar statements to do this.

When you're involved in a feature-length, in-depth story, you'll need much more content, with historical background, context and lots of examples and situations. Here is where you'll need sparklers.

If you're being interviewed for a radio news report and the story is likely to be about 30 seconds in total, you can give the reporter much less material — but it must be from the top of your priority list of core statements. You'll use pillars and supports — but the supports will more likely turn into sound-bites. Reporters love supports.

In kicking off a media campaign, or in developing a personal profile piece for a magazine or newspaper weekend feature or a TV documentary, you can expect to spend hours with investigative-style reporters, and you and the reporter will need far more material. You can't expect to script every word over many hours, nor should you. But you'll need to learn to avoid the traps and say what you intend to say by stressing key points.You'll need to start with a basic series of core statements, in bullet-point form, on one page. We'll teach you how to write it.

"What we have here is a failure to communicate."

Mark Twain was once asked to deliver a speech. He said he would need several days to prepare. He was asked to "ad-lib it." His response? "It takes three days to write a good ad-lib." Many spokespersons can ad-lib, depending on their familiarity with the topic, their subject expertise and readiness to deal with the unexpected questions. But, how often do we wish we had said something better the first time?

It's one thing to ad-lib when the interview is relaxed, the questions are open and everything is proceeding fine. Then, the reporter decides to throw some heat at you. Or finds a subject that you're not comfortable with. Or the reporter decides to undergo a sudden personality change. It has happened often to me and to most of the spokespersons I know.

Some TV reporters start off aggressively as soon as the formal interview starts, catching you off-guard and never letting up. Print reporters tend to move more indirectly toward the jugular, building up to the tough questions after prodding around the subject until they sense some emotional change in your response. For the reporter, it depends on how much time they have with you and how your personalities interact — and mostly how you react when the interview heats up. Meet the reporter from hell.

If you fall victim to reporter traps, if you fail to curb your personality shortfalls, if your reputation vulnerabilities shine through, you've failed to manage the message.

I can remember reading a newspaper article which quoted me saying things I had no recollection of having said. Yet, the reporter had managed to get me to react to certain techniques and the quotes were valid. Not professional, and not retractable, either.

History is littered with the remains of short-lived heroes who experienced a failure to communicate. **Blaming the media for misquotes is the last refuge of a scoundrel.**

With media lines, you'll never finish any communications encounter and then say to yourself:

"If only I had said it this way..."

"I should have stressed..."

"I should not have mentioned..."

"I forgot to highlight..."

"How did I get trapped into saying..."

"Why did my mind go blank on the question about..."

"Why did I use that word?"

"I would have talked about it, except the reporter never asked about change."

"My sentence structure was awkward and I got misquoted."

"Why didn't the reporter use what I said about..."

"What will my boss say when she reads this?"

"I don't remember saying that."

"Have I damaged my organization's relations with that group or person?"

Remember, many spokespersons get into trouble when they try to develop instant statements on topics they're not prepared for or about which they have not given a great deal of thought. That's why one of the bureaucrat's favorite answers is, "we'll have to study that."

A reputation for word-crafting

Good writing, and hence good *MediaSpeak*, results from re-writing or editing. Reporters must meet the requirements of their editors, who have authority to re-write their material, or kill it. Your material must meet the demands of editorial quality. Many large newspapers produce style guides that can help you package your material according to their guidelines. Sentence length, capitalization, leads, hooks, nouns, verbs, subjects, repetition, inverted pyramid style — all tools of editors and spin doctors.

> **"The greatest power in the world is the desire to edit someone else's copy."**
> *— Sign in many newsrooms. It becomes highly edited.*

When spokespersons get into trouble, it's generally because we could have said it better if we had only taken the time to re-strategize, re-write or re-deliver our answer. We sometimes realize, in hindsight, that there was probably a better way to answer a question.

Maybe we could have been a bit more diplomatic or tactful in word or in tone of language.

Maybe we could have used more and better examples to state our case.

Maybe we should have stressed certain elements of the issue that the reporter did not raise by volunteering (or not volunteering) certain information.

Maybe we could have sounded less like a defensive, out of touch bureaucrat and more like a concerned, caring public servant, fine family doctor or civic-minded businessperson.

Maybe we could have come up with a catchy phrase or slogan to represent our position better, rather than talk in PR-Speak or Bureaucrat Speak (BS for short).

Maybe we could have condensed our message and made it clearer to everyone.

Maybe we could have taken advantage of the media interview to change people's opinions or actions, like getting an increase in our budget, approval to add staff or add a new program or product.

Maybe we could have been promoted or fired as a result of what we said.

If only we had prepared a media line and stuck to it. We'd be rich or famous, or at least we would have better communicated what we intended.

One minor slip can result in a damaging situation. I used to have a boss who called media interviews CEO's — career-ending opportunities. I always told him they were career-enhancing opportunities, but I'm not sure he believed me.

You're only as effective as that one slip and all your best efforts can fail with one unintended word. You're only as good as your next gig. Your media lines will protect you and help you — but only if you use them.

The credibility factor

Not all PR people serve as organization spokespersons. Sometimes their job is to assist others. Often there's a sharing of duties between communications staffers and subject matter experts through the designation of a single spokesperson.

So who and what makes you an expert, an instant icon — as today's poster child for your industry, for all hospital administrators, or as the deputy minister of the moment?

Spokespersons don't automatically have credibility — it must be earned. Credibility doesn't come with your title, or as a result of who you work for. Credibility doesn't come with a university degree or as a result of what you may have done in the past.

Credibility is earned and re-earned in every media encounter or customer service encounter. Today's media consumer is more skeptical, more likely to question authority. Your title may not be important. For that brief instant, you are not just the operations manager, you are the company spokesperson.

Choosing the spokesperson is part of the strategy — you choose the person who will best represent your commitment to public-spirited professionalism.

The front-line animal control officer may be a far better spokesperson on pet-care than the politician who chairs the animal-control sub-committee of the committee of parks, etc. The fire prevention officer may be a better spokesperson on some issues than the deputy fire chief. An oil-spill prevention officer may be a better spokesperson on some topics than the CEO. Choosing your organization's spokesperson for each issue is a highly strategic decision.

You gain credibility when you volunteer the right information early in the interview or communications encounter. Sometimes your whole strategy can be conveyed through the examples you use to illustrate what you're doing.

When you manage by walking around, you should gain lots of front-line examples of what's being done to make things better — your "for instances." These are what sales experts call proof statements. We call them sparklers, and they're a key part of your core media message.

When you manage by walking around, you learn the commonly asked questions and answers. You learn to test certain messages outside the boardroom. You get feedback from the front-line. You also get to practice your core messages and improve them.

Our approach, as you'll see again and again, involves equipping the spokesperson with key messages. These core statements will allow you to volunteer interesting information and provide you with credibility. You'll be able to volunteer case studies even when you're not asked directly, and in so doing, take control of the interview agenda and display your credibility.

Comforting the afflicted

There is an old public relations principle that more than anything else, the public want to know that the world's problems are being taken care of by people who care. On the hierarchy of needs, people need comfort. It's been said that the job of a preacher is to comfort the afflicted and to afflict the comfortable.

You make people comfortable when your message is comforting. You can't be comforting unless you appear comfortable — with your message and its delivery.

Your media line is designed to free the spokesperson so that going into an interview you'll know what you are going to say, even if you are strictly limited in how much you **can** say. There may be times when you're not terribly enthusiastic about delivering bad news, but you must be comfortable with the wording of your message.

We offer techniques to distance yourself from bad news when you lack enthusiasm. If you're being paid a salary to serve as a spokesperson, it's your organization's needs that must be met. No matter what personal cause or community issue you represent, you'll still want to display your public-spirited professionalism.

Fill in the blanks

When you're watching a news story about an airplane crash, notice how, except for the details, all stories sound alike: "Investigators from the (blank) agency are combing the wreckage of a (blank) aircraft today in hopes of finding clues to what caused the year's largest aviation disaster. Airline spokesperson Marge Taylor said the plane was a routine flight from…" News stories are highly formula-driven, and your messages can be too.

Many PR departments stock "fill-in-the-blanks" news statements that can be quickly updated for recurring issues. Most police statements follow a formula. A news statement for an oil spill or train derailment contains similarities and should answer common media questions. Crisis management experts know that there are processes to managing an emergency: assessing the situation (the mission), setting priorities (the plan), and assigning responsibilities (the team).

The normal turnaround time for most crisis-related media lines should be two hours at maximum, 30 minutes preferred. This should allow you to develop a short, concise and early message to avoid the phrase "No comment." In some situations, you may have to issue key statements that appear to be spontaneous — when in fact you had planned for such a situation and had an answer ready.

As events occur and coverage develops, you'll want to expand your media line or take steps to focus the media's enquiries elsewhere, to defer or refer the questions and issues. Or you may need to go back to the drawing board and develop all-new messages to meet the changing public opinion environment. Parts of your media line will never change. Other parts will be subject to what's happening in that public opinion environment.

Chapter Four

The skills to succeed

A reputation for plain talk

As a spokesperson, one of your goals is to enter into every interview with a strategy: communicate your media line, and, at the same time, meet the media's needs for short, concise and interesting messages, packaged in plain language. For instance:

Take this book (please). I've chosen a writing style that's often conversational, like a TV script.

There are short words.

Short sentences.

Even grammatical errors!
So what?

Read this material out loud.

This is how many TV writers prepare the scripted words on the TelePrompTer™.

That's how you can write broadcast copy.

It's not like other copy. It's different.

This chapter focuses on three vital communications skills needed to successfully manage every message — your writing skills, your delivery skills and your strategic skills.

Whether you're an individual concerned about neighborhood issues or the president of a large corporation, your basic writing skills will be an essential part of your entire career success. Others will judge you based on your written message — often before they ever meet you.

Good writing results from good re-writing. Learn to "write tight." While it can be easy for someone to take an hour to explain an issue, it can be far tougher to say it in 90 seconds, or five seconds. Yet, that's often what reporters are seeking. The message should come from **you** in a way that meets **their** needs. They're looking for *MediaSpeak*.

Learn how to write from an expert. Take a writing course or engage a writing coach or editor. Read books on writing.

One of the finest writing experts was Dr. Rudolf Flesch, author of *How to write, speak and think more effectively* (Signet and in re-print since it was first published in 1946.) Other books by Flesch include *The art of plain talk* and *The art of readable writing*. While the books are dated, the methods are applicable today.

One of my first bosses, Ev Smallwood of Winnipeg, was a former radio news director. I can remember being called into his office to watch him edit my first draft news release. He was patient and kind, but brutal with a pen. One of his rules was we couldn't use a Thesaurus to find replacement words — they had to come from our vocabulary. I learned that editing can and often should take much longer than writing.

"First of all, let me say..."

If you're going to be issuing news releases, you'll want them to be written much like a news story would be. *MediaSpeak* makes it easier for a reporter or editor to use the news releases and turn them into news stories.

Writing news releases is like writing web pages for the Internet. Use an **"inverted pyramid" style**. This puts the most important information at the top and the details or items of lesser interest can be chopped from the bottom, if necessary, without re-writing the entire news release.

This technique grabs the reader's, listener's or viewer's attention early — in a news release, in a web-page, and in an interview — within five seconds. That's the average attention span of today's male TV viewer.

Most news stories are written (or packaged) this way. Your media lines should be written (or packaged) this way. Assume the reporter has a short attention span — the viewer does.

With the inverted pyramid, **all your key points (pillars) are delivered early**. That way, you'll be able to handle a three minute interview or a three hour interview with the same structure. As you proceed, you add more sparklers, returning repeatedly to your pillar statements. And you can sprinkle support statements throughout.

You can use the inverted pyramid style along with pillars, supports and sparklers when writing *MediaSpeak* — which consists of:

- Your key objective(s), concern(s), goal(s), action(s) or position(s) and these usually come first. We call these **defining pillars**.
- Key questions that need to be answered — your **framing pillars**.
- The major factors in your position, preferably in multiples of three. These can include what you propose to do or are doing, who you're working with or the methodology — what we call "the job, the team and the plan." These are **describing, explaining and informing pillars**.
- Specific examples or applications or "for instances." We call them your **sparklers**. This is where you'll earn your credibility and achieve your communications goals by filling out the back-ground and context. In marketing, these are known as proof statements.

- And you must provide catchy and common phrases that tie everything together. These are your **support statements.**

When you've learned to write in the inverted pyramid style, you'll know how to deliver your media interview in the same way — the first answers contain all your most important points, before you're even asked.

You may have to set aside all of your previous writing and communications methodology when you are writing *MediaSpeak*. Your material is almost always going to be subject to editing that takes place after the interview.

In no other communications situation are you going to be so highly edited, or packaged by the media. You'll want a hand in the packaging.

While a judge or lawyer may have learned how to write a case synopsis that gradually and skillfully builds up to a logical conclusion, a spin doctor may choose to lead with the summation.

You'll increase the "pick-up rate" of your news releases, media lines or statements if they're well written and don't require re-writing by a reporter or editor. Write them in *MediaSpeak*.

If your news releases are not being picked up by the media, it's likely because they're poorly written. It's not always because they are not newsworthy. **Even the most mundane and boring topic can be well-written into an interesting piece of journalism through the magic of good writing.**

Tips to improve your pick-up rate for news releases

- ☞ Avoid clutter. Prepare the release on white letterhead that's clean and plain. Use a minimum number of fonts. Don't get fancy with italics, bold face or underlining.

- ☞ Write an objective-sounding headline that captures or hooks the reader's attention. The first newsroom person to read the release will decide if it goes in the round file.

- ☞ Keep the release to one page only. Aim for a maximum 200 words. Add extra material as backgrounders, if necessary. (Backgrounders are sparklers.)

- ☞ Keep your organization's name out of the first paragraph. The same with your name or that of the person quoted. Consider the issue or topic in the public interest first.

- ☞ Send by facsimile to target editors on your media list. Do not pester them. If the release is not used, re-write it and re-issue it.

- ☞ If you issue your release by e-mail, include the release in the e-mail, not as an attachment. Newsrooms hate to open attachments.

- ☞ Keep paragraphs to four lines or less. Use short words. Write so that when re-formatted, it will look like a newspaper article.

- ☞ Never hyphenate a word at the end of a line.

- ☞ Never split a paragraph, if there's over one page of material.

- ☞ Use a style guide.

- ☞ Always write in the third person, so it can be read, as-is, by a broadcast reporter on-air.

☞ Make sure that the contact person named at the bottom of the news release is:
1.) readily available to answer the telephone immediately — no answering machines.
2.) fully equipped to answer the widest range of questions.
3.) trained by Ian Taylor.

Good writing prevents misquotes

Are you a good writer? If you're not, you may have to hire someone who is or who can teach you. You must enjoy writing and enjoy seeing your work in print or broadcast form. Composition takes many forms — whether we write with a tape recorder, a pen, a keyboard or with our minds as we speak. We're composing messages whenever we speak and good writing skills result in good delivery skills.

It only takes one word to get you into trouble in an interview — it's happened to the best and history is littered with stories of failed media encounters. Often we hear people blaming the media for misquotes, yet many misquotes are avoidable.

Misquotes are often the result of garbage-in, garbage-out. Often the problem would not have occurred if the message had been better prepared and that's what *MediaSpeak* is designed to do.

Misquotes can also occur from poor diction or enunciation — from mumbling or low audibility.

Misquotes can occur from poor sentence structure — where only a part of the sentence is used out of context. You must avoid run-on sentences that are part of some conversational styles. They're hard to edit and can contribute to misquotes.

Misquotes can occur from reporter confusion with long lists of details transposed incorrectly through lack of clarity. Bottom line — improve the message going in and what comes out the other end MAY be improved.

Learn to write like you speak and stop speaking like you write

The key is not to speak like you write, but to write like you speak — unless your goal is to sound like a policy manual or to confine your remarks to legislative or technical jargon.

There are appropriate times to stick to the technical — when quoting the wording of a regulation or law or when explaining highly technical matters. But, in the majority of cases, you'll also want to "translate" technical jargon into plain talk at a grade six comprehension level.

Learn to write, think and speak in conversational style.

Never use a 50-cent word when a five-cent word will do.

Keep your sentences and your written paragraphs short.

Notice how news articles seldom have paragraphs more than four lines long — that's how your news releases should be written.

Learn to write that script for an infomercial — this won't limit you, it will free you to say what you intend to say. You will have considered every part of the issue and if surprises occur in an interview, you'll know how to deal with them because you can use your prepared material.

A reputation for speaking out

No matter how great the message, it has to be *delivered* well, especially if the TV cameras are rolling and your career or your issue, program or office depends on your success.

To be at ease as a spokesperson, you'll have to be an expert public speaker, because when speaking to a reporter, you're communicating through the news media to a potentially large public. You may need presentation skills training and coaching, even if you have lots of experience and consider yourself a natural.

Speaking through the news media is the highest form of public speaking, given the consequences of failure. You might bomb at the Rotary Club, but you'll want to do well when quoted in the financial pages of the *New York Times* when your share prices can be damaged by an ill-chosen phrase.

There are a lot of really good books on public speaking and they cover the corporate and public sector markets. Act soon, while there are still public libraries. Take a voice class. Hire a coach, or two.

Public speaking experience will help reduce nervousness and provide you with experience in handling questions, making detailed preparations and writing skills.

Get voice training. Practice some diction and warm-up exercises for your voice, like these, which you can say out loud before taking to the microphone. Overemphasize the consonants and syllables in each word. Try these out loud somewhere:

We require really weird re-wiring.
Constantly eschew obfuscation.
Double rubber flubber trouble.

One of the largest organizations involved in the development of communications skills is **Toastmasters International**. They have terrific, low-cost training materials and you can proceed at your own pace. Check out a club near you. The annual fees are much lower than you would pay for individual coaching. I'm a charter member of Rainbow Toastmasters, club 4100 in Toronto.

Toastmasters will not only help your speaking skills, but also your skills in meeting management, leadership and much more. In the first year of our club, we saw some terrific improvements in those members who voiced the greatest fears about public speaking.

There is, however, a certain discernible Toastmasters style of speaking that bothers me — it sometimes feels forced and somewhat formulaic, too rehearsed, not always natural. Learn the formulas, by all means, but then take them a step higher and make the methods invisible. The same applies with this book — learn the formulas, then modify them for your own needs.

A reputation for looking right

There's a growth industry in the image-consulting business, however some people in the news should ask for their money back. Image is far more than skin deep.

A former Canadian defense chief was reported to have spent over $20,000 on image consulting. Then he produced a videotape for all staff throughout which he tapped a pen on his desk and shifted his eyes all over the room. The results were predictable — few people believed what he said because of his delivery.

Former Canadian politician Preston Manning went "whole hog" — new hairstyle and color, laser eye surgery, Armani suits and reportedly, voice lessons. His voice still grates like fingernails on a blackboard.

Image is far more than a new hairstyle and ditching the eyeglasses. Image starts with building a reputation for caring and showing that you care about what the public cares about and no amount of packaging can cover up intolerance.

Body language: Neurolinguistics

We all know how important body language can be in any situation. What do you look like on camera? Rent, buy or borrow a camera and find out. Practice your speeches in front of it and play them back, several times. Become your own critic. Take all those "I should haves" and make them "I did its."

There's something magical about what happens between the person being interviewed on TV and the viewer. Within a few seconds, the viewer develops feelings of a direct, personal relationship with the spokesperson. Viewers consider that you are communicating directly and personally to them.

The viewer becomes your customer, your client, your neighbor, your associate, your potential supporter or enemy. And the reporter, no matter how obnoxious or aggressive, becomes a secondary player.

Remember, **it's not the reporter's personality that you need to worry about — it's yours**. The viewer starts to think, "That's how this spokesperson would treat me! This is how much they care about me!"

Your eyes are first in importance, for they're the window into your soul. The viewer will maintain eye contact with you for much longer than occurs in normal face-to-face interaction — up to 100 per cent of the time while you're on screen.

In the vast majority of TV interviews, you will be pictured as a "talking head" speaking to an off-camera reporter to the left or right of the TV set. Yet the viewer's eyes will focus on yours. The slightest eye movement can become a large part of your message.

Some experts say that in normal conversation, we hold eye contact for about seven seconds, look away for seven seconds, then hold again for seven. In a TV interview, you don't know which seven they may use. Practise speaking directly to the camera, then review the results several times.

Your facial expressions, your posture, your hand movements or placement — your entire appearance is part of the message. Learn which hand movements are most effective to your message and which are not. Shifty eyes, nervous eyebrows, facial twitches — all these may be due to nervousness, but how will the viewer respond?

Hand movements should be natural, within a square block below your chin, above the chest and between the shoulders — a small area of space. Keep the palms inward in an embracing gesture that is warm and encompassing. Do not place palms out — it's a defensive gesture of guilt or surrender.

Never have a photo taken with a drink in your hand, even a soft drink. The only exception is the "Clinton," a large Styrofoam coffee cup early in the morning.

Message dressing

What you wear is part of the message you're delivering. You must consider what we call message dressing and it's not complicated. What you wear is part of your strategy. For instance, you may not want people to think you're making too much money, or you may want them to think that you're wildly successful and wealthy. As a spokesperson for an organization or entity, you represent that organization to the public, for however brief a moment.

Determine the stereotypical look or "uniform" for someone in your position or level within your organization. Try to find that happy medium between looking too stereotypical and looking too eccentric. Only a very few people can wear bow-ties and get away with it.

Consider your role in the issue or who has the greatest interest in the issue itself. Consider what the public would expect someone in your position to be wearing. Decide if you need or want to make changes in your appearance. Then, **move toward or away from the stereotype**.

Anyone required to wear a uniform as part of their duties should do so in the interview, including, at times, the official hat. Even a company blazer with your logo becomes a uniform, but bankers wear uniforms, too.

A bank president might not wish to look too stereotypical — no stripes in the blue suit or the tie. Yet certain fashion statements may be too casual or flamboyant. Avoid wearing any outfit that may be out of style in a couple of years — like the 1970s Nehru jacket. A basic navy blue or black blazer has never gone out of style, for men or women.

A business executive might not want to look like a gangster — stick to plain lapels on the suit. A plain shirt or blouse, even a denim one, and a fashionable tie, accessory or scarf are quite acceptable in many on-location interviews where you want to be viewed as a hands-on manager in touch with the issues. A double-breasted suit should be worn buttoned, but make sure it's tailored professionally, and tug down to remove wrinkles or rolls.

Avoid narrow stripes. They're hard to photograph on video because they "buzz" and they also make you look bigger by accentuating any love handles.

Women in male-dominated positions often have to adopt power-styles of dressing without losing femininity. Save your cocktail dresses for interviews held at cocktail parties.

For many women, dark, plain suited outfits work best indoors, lighter pastel colors outdoors, just like Hillary Clinton. Get your fashion inspirations from on-air personalities with figures like yours, remembering that the camera adds twenty percent to your weight and fifty percent to your hair.

A spokesperson for a citizens group should appear as a **serious, qualified and concerned** neighbor, even in casual dress photographed in your neighborhood. A shirt or blouse may be open at the neck, but avoid patterns of any type if you're concerned about your figure. Avoid any shirt without a collar, since the "talking-head" shot shows a lot of neck.

Some of our clergy clients from newer church denominations are instructed always to wear their clergy collars and a black clergy shirt, plain dark suit. The strategy here is to overcome any perception that theirs is not a serious religion or that they're only "playing at church."

Note what others are wearing on TV and adapt the look to yourself. Notice how the TV anchors carefully choose ties matching or contrasting with their jacket and the studio colors.

Wear clothes that flatter or hide your shape, whatever you feel is appropriate.

Shoulders — add padding as required to all outfits. Shoulders frame a talking-head shot. Whenever possible conduct your TV interviews standing up — it improves your posture and can prevent that roll on the back of the jacket under the collar. Never do a sit-down interview in a swivel chair.

Patterns — stick to the one-pattern rule and keep the pattern very small since only a small part of you, the head and shoulders, may appear on camera. For men, that means a patterned tie (avoid stripes), a plain shirt and plain jacket. For women, one pattern only — in the accessories. If you wear a patterned jacket or top, keep the pattern very small, unless you want to look bigger on camera.

Sizes — never tight unless you want to draw attention to your figure. Avoid double-breasted suits which add to your weight on-camera. Always dark for men, no stripes or patterns in the suit. Remember, the camera adds twenty percent to your weight.

Never wear anything that will distract the viewer's attention — unless you want to. No long ear-rings, buttons, bows, frills, pocket puffs, pen or eyeglass holders, fancy jewellery, pocket change or keys.

Yes, this sounds a lot like the instructions given to members of the US Republican Party at a recent convention, or to women candidates of Canada's Liberal Party. The media had a field-day with these training manuals, which were highly ridiculed at the time. Yet all of the on-camera reporters met the dress codes contained in these political handbooks. There will always be exceptions to every rule, to every generalization. But the experts who prepared these instructions have probably learned the hard way.

Inexperienced political candidates can make simple mistakes when they don't know the power of the TV camera with colors, especially reds. Rouge on the cheeks will be magnified.

If you're a champion for the poor and want to wear a t-shirt with a loud political statement on it, go ahead. The resulting photograph will extend that personal message.

If you're spokesperson for the Guelph society for toplessness, make your visual statement as desired.

If you're a police officer, do you really want to look like a 1994 beige Ford LTD with blackwall tires and no hubcaps? Why do so many plainclothes officers choose to look like thugs?

Do all accountants have to look like accountants? All you have to do to soften your look is to change your tie. Smiling helps, too.

But if you're a public affairs officer for the Westhaven Transit System, consider what's expected of you in your role. Keep a dark blazer on hand for those occasional opportunities when you have to be interviewed on camera and your job depends on it. A blazer will not likely go out of style, for men or women. Buy dress shirts without patterns so that you'll always be camera-ready.

A reputation for sincerity — you'll have it made

Bill Clinton could talk all day about healthcare. He's done it many times. Ronald Reagan could talk about healthcare for about seven sound-bites. Reagan was called the great communicator. Go figure who was the actor and who was the politician.

Mark Twain said that once politicians learn to act sincere, they've got it made.

Former Canadian Prime Minister Pierre Trudeau loved to segué into Greek philosophy or existentialism. Prime Minister Chrétien often became very terse with questions, announcing "we have a lot of work to do." There's a lot to be said, however for gaining a reputation for hard work. In his first presidential campaign, Bill Clinton often said that all he wanted to do" was get up in the morning and work hard for the American people." After the Reagan years, this was an appealing concept.

Use pocket cards, if necessary, to remember key points, issues, concerns, facts, questions, examples, names, statistics or media lines. Some people are blessed with phenomenal memories for details, others have to rely on notes. Refer to the notes during the question, or pause after the question, check your notes, then lift your head and start speaking.

In crises or emergency situations, no one is likely to care if you read your material. Most of us are used to seeing the police officer reading from a notebook, as long as they don't overdo the cop-talk.

Some people can memorize a media script or handle an interview by spontaneously delivering great lines. Others have to refer to notes occasionally, speaking extemporaneously. Sometimes we're doing well with our memories, other times we can't remember someone's name. Short term memory is the second thing to go as we age. I can't remember the first.

Acting lessons and on-camera coaching will help with your voice, stance, eye contact and audience connections.

Sometimes we're so nervous we have to read from the written page. If you've never been seen on TV before, people will not expect you to be a star performer and, in fact, some nervousness can win over the viewer who will identify with you. The key is to show that you care.

Shiftiness and defensiveness are negative messages. There are lots of media encounters that must be dealt with by simply reading a prepared news statement of about 200 words, the core messages in a crisis management situation.

Your goal is to avoid having to ad-lib until you learn how to improvise — there's a difference. Once you learn the methods, you'll know how to fast-track your answers.

Can you read from a prepared script and not make it sound like you're reading? Good spokespersons can, just as good speakers can. Some people can read a speech without appearing like they're reading, and that's what you'll have to achieve to be a solid spokesperson. Can you deliver and repeat key messages that are non-committal and purposely vague without sounding like **you're** non-committal and purposely vague, when your job depends on it?

A reputation for understanding differences

If you remember your Psychology 101 classes or basic introduction to customer service or supervisor training, you'll know that there are a wide range of personality types. I'm a control freak with an overbearing disposition and a face designed for radio.

In 1996 I was interviewed for CITY-TV's *MediaTelevision* in their Toronto studio, sitting down, surrounded by TV sets and electronics. The questions were almost all open-style and non-confrontational.

The interview seemed to be lasting a long time because I was "lecturing" the reporter with very long, theoretical and boring classroom-style answers. The reporter asked me why I wasn't answering in sound-bites since that was what I teach. I told her that in order to get me to do that, she needed to put some heat on me, make the questions nastier, make me work. She switched styles and so did I.

By studying personality types we come to a better understanding of the employee-employer relationship, the customer service encounter and the media interview. Reporters all have different personality types and so do we. By learning more about the theory of interactive personalities we can better learn to manage our own and make appropriate changes.

The bottom line on personalities — it's not the reporter's personality that you have to worry about. It's your personality that matters, and how it interacts with the audience. People will look at you in the media and behave as though you are speaking directly to them, and you are.

A reputation for good intentions

Your media line consists of what you intend to say. Casual asides, off-hand remarks, attempts at humor, nervous comments, reactionary statements and reporter-induced wording can become part of your message if you fall for reporter pressure. And some reporters can extract statements you had no intentions of delivering — and they become the quotes.

Note we say "intend" to deliver rather than "hope" or "try" to deliver.

This is one of the areas where your personality type will impact on your interview performance. Some personality types will feel called to answer certain questions in certain ways. Others will learn to avoid the traps in the questions and say what they intend to say.

Brian Mulroney used to say he wanted his cabinet spokespersons to "sing from the same song-book," even though he was not always very successful.

Jean Chrétien, took the practice to new heights. For over 40 years, he insisted on concise briefing notes on issues before him. The notes average about 100 words each, allowing him to communicate in ways that meet the media's need for brevity. He knew he could get into trouble when he strayed from his prepared script. His reputation was seriously damaged when he joked about pepper on his plate, baseball bats and water cannons during news reports on a Vancouver APEC conference.

When we observe politicians in a media scrum, they've often been highly scripted. Their entire strategy may be to deliver only a few concise, direct responses in ways that will assure them 5.2 seconds on the evening news.

Why do politicians use these techniques? For the same reasons that the rest of our clients are trained to use them. To protect their jobs, to display their public-spirited professionalism with credibility and to change or reinforce public opinion.

We've trained political party leaders, municipal politicians and the entire caucus of a provincial political party. We've trained them from the right and left wing and from certain body parts in between. We're far more likely, however, to be training people who are forced, through no fault of their own, to work for politicians. We teach them the same skills we teach all public servants or engineers or office administrators or social activists.

How to manage a scrum

- **Scrum** — a Canadian term meaning a spontaneous news gathering event where reporters surround a spokesperson, often in a lobby or outside an event, like the steps of a courthouse. or at the scene of an emergency.

- **Pacing and timing are everything.** You set the pace with your delivery. You're the chairperson of the event when you take charge.

- **Establish a contract** with the media, letting them know that you have a short statement, to be followed by questions. Wait for three seconds to determine consent through lack of objection. Determine if any reporters are broadcasting live, offer them the first questions.

- **Listen for the topics** contained in the line of questioning and speak to the topics of your choosing.

- **Volunteer information** before you're asked.

- **Avoid concentrating on any one reporter** at a time. Speak to all reporters moving your head from side to side as you speak to all of them.

- **Never run away from the cameras**, unless you want to look guilty. When you run out of material, repeat earlier statements.

A reputation for avoiding becoming the target

Part of a spokesperson's strategy involves choosing the subject of the sentences used in a media line. You have the option of using first person singular, first person plural or third person phrasing. These are highly strategic decisions in your message management.

As a spokesperson for a university study, for example, you have the option of starting an answer at least three different ways:

"My research shows that..."
"Our research shows that..."
"The Alpha-2000 study team found that..."

Or, if you're a police spokesperson:

"My main priority in the investigation will be to..."
"Our main priority in the investigation will be to..."
"Westhaven Police will focus on _______ in the investigation..."

What you're doing here is moving the potential target of the issue from the first person to the third person. When you start answers with "I," guess who can become the perceived issue or problem? There are occasions for the "I" word. They're very specific support statements designed to manage the relationship with the reporter and we'll provide a list later.

When you use the third person (by using the corporate name or "it") you position yourself as the explainer/describer/informer/briefer/reporter rather than the defender. You also help to position yourself and your organization as **part of the solution rather than the problem.**

Use of the word "I" relates to your level within an organization and your authority to make commitments on behalf of that organization. You can also be seen as egotistical and self-serving when you talk about "my study, my goal, my main priority." **There's no "I" in team.**

The use of "we" or "our" can change the way the public will perceive your message. You will also be portraying yourself as a team member. And when you can speak in the third person, you've transferred the focus in a highly strategic way.

Using your corporate or organization name or your web site provides publicity and brand recognition through added mentions — even when you have a company name like **Never Say "NO**

COMMENT" Incorporated — which can become a bit cumbersome and takes over two seconds to say, but few forget it.

Sometimes spokespersons move back and forth between the use of the first person plural (we) and the third person (it). This can be purposeful and strategic, even within an organization's many layers. For instance, do you wish to position yourself as a police spokesperson, a Westhaven Police Force spokesperson, a spokesperson for the homicide squad, for the chief of police, for the communications branch, or for the division commander?

Are you a spokesperson for the "Government," your individual department, or for the department's planning office within the south-west region of your state or province? Each role provides a different perspective in the interview and can protect you from, or expose you to larger questions. Do you want to position yourself as a spokesperson for the Minister? Are you a political staff member? Or are you a spokesperson for the construction branch who will refer policy questions elsewhere?

It may be an important tactical move to position yourself very carefully to avoid being seen as the center or cause of the controversy. You'll likely want to be seen as part of the solution to it. Or you may purposely change positions throughout the interview to serve various purposes.

Your strategy may be to buy time, to re-direct the issue, to position yourself or your organization as a "victim," to raise new issues or to place the greater issue in context. Your strategy may be to illustrate the need for more staff, more resources or more action without criticizing the hand that feeds you. Your strategy may be to communicate the fact that you only have three staff to handle 3,000 enquiries. The key may be to avoid the use of the word "only."

A reputation for getting more resources

One of the most important issues today, for many of my clients, is staff reductions and layoffs. This is an issue in government, business and elsewhere. Downsizing is the buzzword we sometimes hate.

If you say you don't have enough staff, you may be seen as critical of your department's staffing decisions. If you say you have enough staff, you may not feel as though you're being truthful, especially when you're trying to get more staff. Yet your goal may be to convince people that you need more resources. So, prove it diplomatically. Show why you need more ambulances or drivers or inspectors by using the facts.

Use quantitative and factual, pillar-type answers:

> **"We have four inspectors to cover an area of 32,600 square kilometers and inspect 1,643 facilities each year. We're working to make their jobs more effective by:**
> > **building relationships with...,**
> > **developing new...,**
> > **and improving...**
> > **The key to the inspection process is...**
> > **We're constantly looking at new ways to improve efficiency."**

The implication here is that you're short-staffed for the job required of you, but you haven't said so directly. Explain how much work needs doing and how much is being done. Never say "We're doing the best we can with the resources available." That's a cop-out. Are you trying to draw attention to the lack of resources? Instead, say, "We're doing this, this and this."

If there are delays or service deficiencies, explain them. Don't get defensive. It won't take the reporter very long to find someone who'll say you need more staff. Your facts will contribute to "ventilating" the issue so that there's no need for you to express what might be viewed as an unofficial opinion or comment.

Chapter Five

Be bold, or be toast

With *MediaSpeak,* it's the answers, stupid — no matter what you're asked

In our intensive *MediaSpeak* workshops, we first train students to manage an edited-style interview in which the reporter is seeking one or two short quotes. Once you've learned to manage your message in front of a hostile reporter, you can modify the techniques to a wide range of communications formats. The key, of course, is how you answer the first question.

The first, essential technique is to establish control of the agenda and define the issue with your first pillar answer. Then, offer the reporter a specific example (a sparkler) of what is being done (or should be done) about this issue. To add emotion or empathy, use a support statement or two. This is a classroom formula which, once learned, can be later modified to meet your needs.

Your personality type will determine how comfortable you are with using a formula, but it works for the experts, and you'll notice how similar their methods are to those we discuss here. Former Ontario Premier Mike "the knife" Harris often used a power pillar, and no one has ever accused the Premier of being an intellectual. He uses an easy-to-learn technique.

This training technique is based on the assumption that you'll only get one chance to say what you intend to say and what you will be quoted saying. Having decided what you intend to say, it will be your objective to say this first and probably to say it again.

Media interviews are not like job interviews, where you have a certain amount of time, where there are several opportunities to display your knowledge and where you'll often be given a chance to provide information that you were not asked about. Like the old advertising slogan, **you never get a second chance to make a first impression**.

There's something almost mystical, or mythical, about what happens between you and the reporter during the first question and first answer in a media interview, especially when the TV camera is rolling and your career is on the line. This is the same time that you're most nervous and reporters know it. Some will work to make you more nervous, or defensive, or angry.

How you handle your first answer speaks volumes to the reporter about your preparedness, your knowledge, your self-confidence and your ability to communicate. If you falter or flub an answer early in the interview, it may not matter how well you do later because the damage has been done.

If the reporter has succeeded in getting you to say something a certain way, nothing else that's said can replace it. You will be quoted saying it the way the reporter has trapped you into saying it.

If you take control of the agenda early in the interview, you'll be sending the reporter powerful signals about control. And control is really what it's all about.

If you've ever taken personality tests you'll know that some people are naturally more controlling, assertive or manipulative. Other personality types may not easily lend themselves to some of the techniques we propose. But the experts use these techniques and you can too.

What the experts do is blend their assertiveness skills with their empathy skills in stress management situations. Or, caring, speaking out and action skills. And at no time are they more on display than a media encounter. You'll need skills in negotiating, assertiveness, customer service and crisis management, as well as the presentation and delivery skills I spoke about earlier.

How will you take control early in the interview? By using a bridging technique which is as simple as a phrase like "before I answer that question," or "let me explain our concerns, plans, situation, approach, method or objectives."

The reporter sees that you're offering a package of concise, direct and interesting information, despite what you're asked. To illustrate your product knowledge, your MBWA skills or the seriousness of the issue, the bridging technique can start with a phrase like, "Let me give you an example."

You can't always make an issue or a question go away. You can deflect it three times in an average interview. If the reporter still returns to the line of questioning, you have to address it directly, even if your acknowledgment is as short as, "Yes, that is a concern. Our greater concern is…" Never let anyone else out-concern you.

Want to practice your control techniques? I look forward to calls from telemarketing firms, especially those representing stock brokers trying to send or sell me something. I've learned to turn the call into a marketing opportunity for myself.

I sometimes ask for the name and telephone number of the President of the company calling me. Or I ask for the name of the person who handles public relations.

Then, I ask to have the President give me a call so that I can talk about my services. The key is to take control of the telephone call very early and take over the agenda. It hasn't helped business yet, but it gives me something to do on quiet, creative days.

Case Study: Caring about your employees — Hospital layoffs

Say the interview is about recent hospital layoffs. The TV reporter has the benefit of a camera and lights glaring at you and the microphone is suddenly thrust at you after this multiple and emotional question. Meet the "reporter from hell."

> *"How about the families of laid off workers? How many of them will have to go on welfare while fat-cat doctors and administrators pull in massive salaries? Can you sleep at night knowing the devastation these layoffs will create?"*

Is the question about your sleeping habits? Is it about fat-cat doctors? Is it about your salary? It can be, if you permit it to be by responding directly to those parts of the questions. Or, you can decide to interpret the question as an opportunity to talk about how much you care about your staff, and the efforts your hospital administration made to keep them in the face of government cutbacks. Your goal is to deflect part of the issue and position your hospital as a victim — without using the word "victim."

Here's a fairly complete media line for this interview topic:

> *"Let me explain what's happening today. (Pause two seconds. You're taking control of the agenda with a bridging phrase.)*
>
> *"Today's announced layoffs at Westhaven Centenary Hospital result from a massive re-structuring of health delivery services due to government cutbacks. (You're staking out your role and defining the issue with a pillar statement.)*
>
> *"The re-structuring team has identified the need for a 22% staff reduction. This will affect some long-term, loyal and dedicated employees who are part of our hospital family and we're sorry to lose each and every one of them. (You're positioning yourself on the emotional side of the issue.)*

"The decision to reduce staff was one of the toughest ever made by hospital administrators and board members. Before reducing staff, we worked to achieve other savings, such as…" (Identify three examples, or sparklers.)

"Our goal has always been to expand and enhance health services in growing residential neighborhoods, including…" (List highlights of at least three such initiatives here. Indicate their status as a result of reductions. You're staking out these strategic examples as programs worth saving.)

"The alternative, however, was to face complete closure. Some of the other ways we'll achieve further savings will be through…"(Give three extensive and specific examples using varying degrees of technical language and "hospital-speak." You're displaying your expertise and reflecting on the hospital's public-spirited professionalism.)

What you're doing here is…

1. Allowing the question to serve as a prompt or opportunity to deliver your media line.

2. Taking control of the agenda with your first answer.

3. Volunteering specific examples that illustrate your subject expertise.

4. Positioning your organization as a victim of provincial cuts.

5. Sticking to your media line.

6. Avoiding traps in the question.

7. Delivering much of your message in one package without the questions controlling the message.

There are over a dozen sound-bites in this answer — more than enough for almost any reporter. You'll probably deliver several statements, then wait for the next question. Then move on with more of your media line.

Have you ignored the questions? You've dealt with the issue of layoffs and of how layoffs are impacting your facility and those who made the decisions. You've ignored the traps in the question. You've re-defined the opportunity in the question to deliver your message.

Have you avoided the tough issues in the question? You've dealt directly with the key parts of the issue — why there were layoffs and how management feels about them. You've also volunteered specific information on program initiatives that are specific and probably well-known. You've given the reporter interesting and helpful material for a story.

Have you taken control of the interview? Yes. You've also answered specific questions that have not yet been asked, taking control of the agenda and volunteering material the reporter can use.

Are you behaving like the experts? Yes, this is exactly how they manage interviews. Sure it can be frustrating to see some spokes-persons refusing to answer a question — it happens at least once a day on the news. It's frustrating because they've handled the interview poorly. They have not had spontaneous responses ready. Responses that combine that right proportion of empathy and assertiveness, along with specific examples to establish credibility and prove their positions.

A reputation for finding opportunities

Skilled, experienced spokespersons learn to listen for opportunities. Any question is an opportunity for you to talk about anything you want, by building linkages, bridging to your topic or firmly taking control of the communications agenda. The key is to do it spontaneously, assertively and empathetically, with all the crisis management skills necessary.

When you read a quote in a newspaper, do you have any idea what the question was? When you see a talking head on a TV news story, do you know what words were used to elicit that response? Only when the interview is live do we hear the full questions and the full answers without cutting, pasting and editing.

We actually hear a reporter's question in fewer than five percent of the media interviews in daily news coverage. Under five percent, and those are always radio or TV. It's not the question that gets you into trouble, it's the answer. Your media line contains the answers.

You may have to learn all new ways of listening to the question based on this process:

☞ What is the question about? Or, what do I want the issue to be about?

☞ What do I intend to say on that topic?

☞ How do I connect my answer to the question while avoiding traps or inflammatory language?

☞ How do I appear spontaneous and part of the solution rather than the problem?

By sticking to your media line.

Start with one word or phrase

The easiest way to start your media line is with one word or phrase. Attach it to one of the issue-defining phrases later in this chapter.

The experts agree that this should be a word that communicates in inarguable, least arguable or inalienable terms. Talk about your tongue-twisters.

Words like: *safety, community, the future, caring, helping, sharing, concern, building, developing, justice, fairness, equality, responsibility, duty, accountability, thorough, detailed, efficient, working, co-operating, improving and comprehensive.*

Direct and indirect quotes

An indirect quote is one where the reporter writes: "The Prime Minister said ***that it's*** damn the torpedoes, full-steam ahead for the new regulations, despite objections from groups like…"

Indirect quotes account for over half of what we see in the media and they allow reporters to take immense liberties in communicating what spokespersons say. In some newspaper stories, there are more indirect quotes than direct quotes — those that contain the actual words used — in or out of context. Read through your favorite daily newspaper and observe the number of direct quotes — in quotation marks; and indirect quotes — not in quotation marks.

The reporter is taking a lot of liberties here — something called journalistic licence, although no one issues these licences. But if our answers had been better, we would be quoted saying what we had really said in a way that met the media's requirements for brevity, plain talk and interest.

Are we teaching our students the sleazy techniques used by politicians not to answer the question?

> "Our goal is to assist our students to display their public-spirited professionalism. We do this by equipping them with the skills to…" (pillars.)

> "Let me give you an example of a challenge faced by a recent student. In this case…" (Sparkler.)

A reputation for openness

You can carry a lot of old or recent baggage into an interview. You probably don't want to talk about it, but it won't go away. The old advice applies: put it behind you and move on.

When you know they're going to rub your nose in it, deal with the issue with your head held high and move on. Find some way of stating your message by saying things like:

"Yes, our critics were right."

"People were very upset at the decision."

"We made a mistake."

"We failed to study…"

"We moved too quickly."

"We should have…"

"We learned from the experience."

"We're taking steps to make sure it doesn't happen again."

Political correctness is all about not offending people.

Some phrases or words commonly used in the past are no longer acceptable. If people are hurt, apologize publicly, and if necessary, repeatedly. If you have to grovel, do so. Apologize until you're blue in the face, if that's what it takes. But have an action plan ready to stress. Let the apology serve as a lead-in to your remedies.

An apology can be as direct as you make it. You can say you're sorry for having said something, or you can say you're sorry that people were offended by what you said — there's a big difference.

It may be a tongue-in-cheek apology, common in the Parliamentary system, where you apologize for saying that the minister was lying. That you did not intend to call the minister a liar, and that you hope no one will consider that the minister is a big, fat liar based on what you've said. Chances are the minister will wish you'd just shut up.

Self-deprecation

While the use of humor is always dangerous, an exception is self-deprecation — making fun of yourself. Since you're the recipient of the humor, you're the only one likely to be offended.

As Toronto's most famous merchant, Honest Ed Mirvish, often said, "I'm full of humbility."

Part Two

Writing and Delivering MediaSpeak

Chapter Six

Pillars

A note to readers:

I'm going to change my writing style for a moment and start this chapter with point-form highlights, statements which serve as briefing points. This is how I write speech notes and media line pillars, and you must learn the same. So, let's get right to the point.

Highlights

What are Pillars?

Pillars are short, concise and direct statements that:

- ☞ define or re-define the issue in inarguable or least-arguable language or in inalienable terms.
- ☞ state, define or position your role, mission or concern.
- ☞ define your job, expertise, function responsibility, task(s).
- ☞ frame or re-frame the question in the public's mind.
- ☞ outline main reasons, factors, statistics, actions, concerns, challenges, steps, teams, plans, partners.
- ☞ inform, describe and explain — rather than defend.

What do they do for you?

☞ They prepare you for the toughest questions.

☞ They control the interview process by anchoring and sectioning your presentation.

☞ They're designed for repetition and combining into power pillars.

☞ They're for use with supports and sparklers.

☞ They keep the reporter on track.

☞ They're the bare minimum for any interview.

The results?

☞ Provide appearance of spontaneity.

☞ You'll never have to ad-lib.

☞ You can offer new insights, approaches or aspects of the issue without being asked.

☞ You can create linkages between other issues.

☞ Buy time until decisions are made.

☞ Deflect the issue without saying "not my department."

☞ Raise new concerns or information.

☞ Position you as part of the solution.

Technical and writing:

☞ Start with short answers to a list of open-style questions:

- What's your main concern?
- Why are you concerned?
- Describe the situation or circumstances.
- What is the impact on the public?
- What is being or should be done?
- How will it be done?
- What are the main facts and main factors?
- Who will you work with?
- When will it be done?
- How will the public benefit?

- ☞ Move to three point answers for each of the above questions.
- ☞ Predict the toughest questions and answer them in under 15 words, with existing answers.
- ☞ Must be plain language — grade six level.
- ☞ No jargon or buzzwords.
- ☞ Write in bullet points, like this list.

Meet media's requirements for:

- ☞ Definition of issue, concern.
- ☞ Brevity.
- ☞ Clarity.
- ☞ Create power sound-bites.

A reputation for getting to the point

If you're a police officer and the topic of the interview is home security, there may be a number of issues or sub-issues that you must be prepared for, even if only to refer the questions elsewhere or defer them until you know the answer.

Home security may raise other issues such as: private security firms, Neighborhood Watch, burglar alarms, statistics on crime rates, statistics based on race or young offenders, canine protection, police helicopters, juvenile crime rates, staffing levels of your police force, political concerns from municipal, provincial or federal governments and recent high-profile incidents that are in the news. What is it, specifically, that reporters are seeking from you? Ask them.

The issue of police staffing levels, if you agree to talk about it, may raise questions about employment equity, hiring practices, minority rights, pensions, age discrimination, municipal taxes, city hall politics, the chief's personality, provincial cutbacks, prison sentencing, parole and more. You can expect the reporter might ask you about any high-profile police issue in the news this morning, no matter what your involvement.

Are you prepared for all of those sub-issues, or will it be necessary to refer them to other spokespersons? You must expect them to arise, no matter how carefully you negotiate the interview. Once the TV camera is rolling, all your best intentions can change under the pressures.

It can be much harder, under pressure, to refer questions to those with more expertise or who are better able to answer them. It can be a bit easier when you're being interviewed by telephone, but under the pressure of the camera or in the studio, the reporter from hell has certain advantages — and you must expect the reporter to push you where you may not choose to go.

If you're a spokesperson for a hospital, the topic may be health care delivery, but there are dozens of issues and sub-issues: length of hospital stays, homecare, age of patients, maternity care, funds recently spent on other projects — like the new statue in the lobby or the salary of your hospital administrator, union/management concerns, recent layoffs of nurses, ambulance services, transportation for the disabled, and government funding levels. There are issues which relate to professional associations, politics and neighborhood relations.

If you're a spokesperson for the United Way, there are issues of faith, ethics or "morality" (abortion, right-to-die or visitation rights under power-of-attorney) and politics (workfare). Even if the reporter tells you the interview may only be about your new fund-raising campaign, each of these issues may arise and you must be prepared to deal with them. The reporter may argue, convincingly, that all of these sub-issues do, in fact, deal with fund-raising. To this, you can respond "Yes, fund-raising is part of the issue, that's why we're examining it. Our main concern is…"

A good media line will have sample statements that will allow you to refer the questions elsewhere, to defer answering the specific question until your position is developed or to re-direct the issue to your agenda with a phrase like: "The real issue is…"

That's what your media line is designed to achieve — it must equip you with spontaneous answers to almost any question that may arise. Or to re-define the issue to one of your choosing.

How PR people overcome their reputation

It's been said that **the main function of a public affairs expert** is to help your organization identify those issues (or, more specifically, the tough questions) that may arise so that you will be able, at some stage, to deal with these issues spontaneously and professionally, with empathy and assertiveness in proper balance.

A second function of public affairs experts is to define or re-define the issues in a concise, tactful and public-spirited way. Sometimes you're so close to the issue, or you've been bombarded with information and criticism for so long, or you've only been involved from one perspective — and you can't see the forest for the trees.

This is where an outside perspective can be so helpful.

This is where some media consultants can step in and examine the issue from the "outside" or from all sides without being defensive or reactionary.

The role of the spokesperson is to explain and describe an issue or topic in a way that reflects positively on the public-spirited professionalism of the organization you serve, yet you may not have had time to synthesize the message into brief sound-bites.

Say your hospital has canceled certain maternity programs or services. Are you abdicating your responsibilities to mothers and future generations of children? Or, are you re-focusing your priorities to match changing demand for healthcare related to demographic changes in society? This is an example of spin.

Spokespersons must often avoid being placed in defensive or reactionary positions that are argumentative or inflammatory. Your goal may be to attack the government for its action, but how will you conduct yourself if the attack is aimed at your organization? Will the attack become the issue or will you be able to deflect it back at the original target?

Media lines are often required because of the way that an issue is currently defined by others — your critics, editorials, columnists or the "spin" or approach that front-line reporters and re-write editors have taken in covering the story to date. Any news story can be approached by a wide range of angles or spin.

Competing news agencies may cover the same story in widely different ways or they may all accept, without question, the "spin" that has been applied by one party to the issue. When that spin has been well-prepared and meets the media's needs as concise and catchy, it may not be questioned or challenged. Or it can stand up to questions and challenges because of the expert craftsmanship of its writing.

Case Study: Still caring about safety — Photo radar

In the early 1990s, Ontario introduced photo radar to catch highway speeders. Critics of the program defined the issue in several ways: a "cash-grab," an invasion of privacy, a failure to provide adequate numbers of police officers, a reluctance to increase speed limits and an overzealous enforcement approach to speeding.

Yet in working with the photo radar project team, we constantly reminded spokespersons to define the issue as one of road safety. And we developed core statements in class, many of which were used extensively in dozens of interviews later. These statements included use of accident statistics from other jurisdictions as well as technical aspects of the radar equipment.

Who can argue with safety? Or with protecting the environment, or with improving customer service, or with wanting to hear the views of all concerned? These represent inarguable premises.

*Who represents the interests of speeders and lawbreakers? Who can refute or argue against statistics when they show declining fatality rates? There's another old public relations adage that says, "**get on the right side of the issue**." Your media lines should be designed to achieve that goal.*

Case Study: Caring about customers — Airport taxis

Several years ago, there was a situation at Pearson Airport that symbolized the need to develop a media line and to stick to it. The media line had been approved by senior management and as the spokesperson, I realized that my job depended on handling the interview professionally — I had only just arrived in Toronto.

Taxi and limousine drivers were enraged over recent changes to the contract for bus service between the airport and additional downtown hotels. There were demonstrations and picket lines surrounding the airport administration building, providing a very public media photo opportunity. There was no way I could or would say "No comment."

The taxi and limousine operators had four demands of Pearson Airport administrators:

- *Cancellation of the new bus contract;*
- *Reductions in their licence fees to compensate for lower revenues (allegedly) in their business;*
- *A royal commission of enquiry into ground transportation at Pearson Airport;*
- *The firing of the airport general manager.*

In a TV interview, conducted in front of Toronto's Royal York Hotel, the first reporter question was one of the worst I could imagine:

- ***Question:** "Are you going to fire the airport general manager in response to demands from the taxi and limousine industry?"*
- ***Answer:** "There's really only one issue here and that's improving ground transportation services for air travelers. This new bus contract will provide better service and more travel options for air travelers. Transport Canada has no intention of changing our commitment to improving customer service."*

The reporter said, to my immense and everlasting surprise:
"Thank you very much, that's all we need."

The quote ran, as delivered, on the evening's news. **This is a 43-word statement, about 20 seconds.** *It contains three pillar statements.*

I delivered the media line in response to a question that dealt with the most sensitive aspect of the issue. I hadn't really answered the direct focus of the question. But it seemed that it sent some message to the reporter that I was not going to be drawn into her agenda. And I was shocked at how well it worked.

What if the reporter had repeated the exact same question? The answer would have been (in hindsight, of course) much the same. Repetition is an essential technique to take control of the interview, especially an interview which will be edited for later use.

Had the reporter asked a third time about firing the airport manager, then and only then should you deal directly with the question. Then, you would point out that the question relates to one of privacy and you're not authorized to discuss issues of privacy. You would, on the third question, define the issue in the question as one of privacy.

What if the reporter kept coming back to the issue of the role of the airport manager in the new bus contract. The answer might have been, "Over a dozen employees were involved in the contract and each of them had the same objective, to improve customer service."

In our workshops and training courses, I've used that example hundreds of times to illustrate the importance of media lines. That example is one my favorite, personal sparklers.

Did I want to talk about firing the airport manager? Spokespersons are not authorized to discuss personnel issues. The real issue in the interview, was, according to what I said, improving customer service. Who can argue with that? It's an inarguable premise.

> **There's an old public relations adage that states, "If you hold up as your position, an inarguable premise, your objectives are more than halfway achieved."**
>
> ***—The law of the inarguable premise.***

Did I want the reporter or our critics to establish the agenda for the issue? Not if I could avoid it. And you won't be able to avoid reporter traps unless you use certain techniques, just like the experts.

You'll never be able to avoid an issue that calls out to be dealt with. *Others will quickly step in to fill the communications vacuum. But you can work to re-define the issue while avoiding traps in certain questions. And there is always a way to communicate something so that you display your public-spirited professionalism.*

Case Study: Caring about the environment — Aircraft noise and neighbors

Once upon a time, a community information meeting was being held in Malton, Ontario to deal with increasing complaints from residents about aircraft noise near Pearson Airport. It was a very volatile and angry meeting at which people were justifiably upset about increased noise that resulted from the summer's construction program. The air traffic controllers had to direct all air traffic over one Malton neighborhood. They weren't happy campers.

Our media strategy included putting the airport "on the right side of the issue."

During an interview, with then-CBC reporter Steve Paikin, the first question was, "So I guess you bureaucrats got a pretty rough ride tonight from the people of Malton?"

The answer:

"Area residents have been experiencing record levels of noise

this summer and we apologize for the environmental impact that has occurred." (Apology Pillar.)

"Some of these residents have aircraft flying over their homes every two minutes, 18 hours a day and we're anxious to reduce that." (Power pillar.)

"Our goal is to be a responsible neighbor." (Support)

"We're working on a new noise management policy. It will address air traffic control procedures, hours of service for certain quiet aircraft and improved aircraft design by the airlines operating at Pearson." (Pillars.)

"Aircraft noise is an international issue requiring the co-operation of the airlines, airport operators and neighboring municipalities. Many of our efforts at Pearson are similar to those at Chicago O'Hare, London's Heathrow and JFK Airport in New York." (Sparkler.)

"Pearson Airport's construction team is working to assure that runway repairs are completed as soon as possible. We're working to resume our normal runway rotation program." (Pillars.)

The reporter asked his question again, and a very similar answer was given the second time, to which the reporter remarked later on the spokesperson's "smoothness."

Smoothness had absolutely nothing to do with it. That delivery technique was motivated by fear of losing my job. I'd prepared a media line and my job required me to deliver it. I wasn't always so successful, of course, otherwise I wouldn't have become a consultant. There have been dozens of occasions when I wished I'd developed and stuck to a media line in every media encounter — no matter how brief or informal the media line.

Was there a lot of nervousness involved in such an interview? Of course. But not as much nervousness as when I gave an ad-lib interview and then had to wait 24 hours to see what appeared in the next day's newspaper because I couldn't remember everything I said.

Remember, your boss probably reads the paper to start the day.

Using media lines is the cornerstone of media interview skills management: ***You must take control of the communications agenda.*** *You must send a strong signal to the reporter that there are core statements that you intend to deliver — no matter what you're asked. These are the statements you intend to say and that you can then be quoted as saying.*

You cannot ignore *a topic or an issue, but you often must re-define the issue or avoid using certain glaring or inflammatory words that the reporter places in the question or words that your critics may be using.*

Chapter Seven

Reputation words and the "not" word

> **"I am not a crook."**
> *— Richard Nixon, professional crook*

Recently, the head of an electronics firm was quoted about a piece of equipment that failed, throwing Canada's telecommunications systems into a panic. His answer, "Well, I wouldn't exactly use the word lemon."

What was the question? The question likely (we don't know because it was edited out of the newscast) contained the word lemon. The reporter trapped the spokesperson into using the word lemon. Instant quote. It's a technique called negative questioning and reporters use it far more than most care to admit.

Check out today's newspaper or tonight's TV news and count the "nots." About a third of the quotes contain the word not, or have headlines like **"Company denies price-gouging"**.

In some political worlds there's an old and still practiced method to win elections. You simply accuse your opponent of having had carnal relations with barnyard animals and pray that they spend the rest of the campaign vehemently denying it while you promise to build roads and/or reduce taxes.

Today, all you have to do is accuse the local police department of racism and one of them is sure to go ballistic. The more they protest, the more the public feels there's some justification to the attack.

Complete this statement: "I/we want to build a reputation for ______________."

Now, replace that word with its opposite: "I/we want to avoid having a reputation for ______________."

These are called reputation vulnerabilities, a euphemism if I ever heard one. You will not likely want to speak these words aloud in an interview, of course.

In planning your messages, wouldn't you want to build a reputation for plain talk, candor and hard work?

How are others defining the issue? How is the reporter defining the issue? Does it need to be re-defined? That's what your first pillar statements must do.

Chances are, others will be defining the issue in more inflammatory language. Or, maybe your goal is to attack, using attack language. The strategy may be to get another spokesperson to deny the word.

Negative questioning

One of a reporter's favorite traps is the denial or negative question.

> *Did you smoke marijuana, Mr. Politician? One of them said he had been at parties where marijuana was present, but he preferred corned beef. He did not deny it, but the issue went away.*

Canadian author Mordecai Richler once wrote about getting a phone call from a reporter and being asked if he used illegal drugs. Richler said he didn't. The next day's headline read "Richler denies drug addiction".

There are occasions when your goal is to deny certain words. Often, however, you can be trapped into using them and appearing defensive. The secret is to replace these words with ones of your choice. I avoid referring to the replacements as "positive" words. I prefer to call them public-spirited and professional words.

"Not" is a dangerous word

Your success may depend on your weakest answer, one ill-spoken word, even if the word is delivered in the negative sense.

Often, when we see ourselves quoted using glaring or inflammatory language, it's because the reporter trapped us into using that language based on the words in the question — the loaded question creates a denial or "not" answer.

It's one thing to use the "not" word purposely. It's something else when the reporter traps you into using it for a quote that's unplanned.

It's almost always preferable to say what you are doing rather than what you're not doing — when the not is followed by an inflammatory word or phrase, like "we're not covering up."

I've never once felt I've been misquoted, or burned, when I stuck to my media line. It's only when I ad-libbed and was unhappy with what I said that I wanted to blame my problems on the reporter.

Had I not said what I said, I would not have been quoted saying it. Had I said what I should have said, I would have been quoted saying it. Whenever I've stuck to my media line, without "winging it," I've surprised myself with the results. Whenever I've gotten into trouble, it's because I didn't plan my answers well enough.

How many times in our personal lives do we wish we had not said something? Or that we had thought a bit more before saying the first thing that pops into our heads? It's one thing to say something stupid to our spouses — most times, they can be forgiving.

Saying something stupid on the front page of the daily newspaper can be bad for business.

Reporters seek spontaneity as a reflection of our true feelings, and that's fine when we're the person on the street talking about broad social issues. However, we train professional spokespersons.

If a reporter were to ask us if we were teaching you a lot of "sleazy methods not to answer questions," we could reply by saying "in a manner of speaking, I suppose." The answer means "yes." Then the reporter could quote us saying so, even if it were done **indirectly**.

Even if we answer using the word "no," the reporter can still quote us using the term "sleazy methods" in an indirect quote. Or the reporter could use journalistic licence and say, "Taylor denies teaching the sleazy methods politicians use every day." Taylor didn't say it directly. But the word "no" is an indirect response and can generate an indirect quote.

I have a list of words that reporters love. They're attack words. They can result in answers that contain the word "not." It's one of the most interesting words in our language. In the art of *MediaSpeak,* however, you must be very careful about describing what you are not. Instead, say what you are, in words of your choosing.

This list is not intended to be complete, but it should get you started. Here's how to use it. Pick a glaring word and create an attack question with it. Like, "Why are you abdicating your responsibilities to healthcare/ education/ safety/ our future generations?" Do not repeat the word "abdicate." Replace it with an inarguable word, like change, "We're changing the way we deliver healthcare."

Sample questions:

"Is there a cover-up?"

"Are you playing fast and loose with the figures?"

"Are you racist?"

"Are you a crook?"

Do this with several words. **Every glaring, accusatory or ill-defined word is an opportunity for you to replace it. Do not repeat it unless you want it used**. Or, you can use these words to attack your opponents, but only if you want to invite attacks from others.

Remember, there are times when these words will be part of your media line — but only when they're planned or when you want others to deny them. But consider what happens when **you** deny them. You're then quoted using them. Someone else wins at the game of spin.

Yes, these are often name-calling and sometimes in poor taste. But if you're ready for them, you're ready for any question. These are some of my favorite words for attack questions in our training workshops.

Opportunity words from the "reporter from hell"

A abandon, abdicate, abuse, acrimonious, ageist, allegations, angry, anti-something, appalling, assassinate

B baby-killer, back-stabber, benevolent dictator, bible baseball, bible-thumping, bigot, blame, blunder, blur, bomb, boondoggle, botch, breeder, bulldoze, bum-boy, bungle, bureaucrat

C carpet-bagger, card-carrying, character assassin, charlatan, cheat, circle the wagons, closet-something, coat-tails, collusion, common sense, compulsory, con-artist, condescending, controlling, copy-cat, corrupt, cover-up, crazy, crook (see also Richard Nixon)

D dangerous, dangling, decay, defensive, destroy, disaster, disturb, domineering, drag feet, dreadful, drop the ball, drunk with power

E eco-terrorist, eccentric, egotistic, elitist, embarrassed, enough (as in doing enough?), environmental disaster, enviro-nazi, evil-doer, expert — so-called, exploit

F family values — without valuing families, fast and loose, feminazi, feminist, fiddle while city burns, flak, flimflam, flip-flop, frantic, fudge the numbers, fundamentalist

G something-gate, goodie-two-shoes, go round in circles, grey-haired, gloss over, gross stupidity, guru

H hand-out, hard-luck, hard-up, hero-worship, heterosexist, hidden agenda, hide behind (as in badge), hide from (as in facts), high flyer, hocus pocus, homophobic

I idiotic, ignore needs of, ill-advised, illegal, illegitimate, illogical, Imelda Marcos, in bed with…, incestuous, inciteful, incompetent, incomprehensive, incongruous, incriminating, indecisive, insensitive, insubordinate, insufficient, irresponsible

J job-creating exercise (see bureaucrats), juggling

K kow-tow (see racist)

L lackadaisical, lackey, lacklustre, languish, lemon, let-down, lie, lifestyle, lookist, loose lips, loose with facts, lord it over, lose face, lose control, lost hope, lost cause

M made-up, martyr, mealy-mouthed, mean-spirited, mega-something, mess, militant, misdirected, miserable, mish-mash, misguided, mislead, mismanage, monopolistic, morale, Mother Theresa, mumbo-jumbo

N nay-sayer, nazi, nepotism, nightmare

O obfuscate, old, oligarchy, oppose, outlandish, out of bounds, out of date, out of ideas, out of touch, over-compensate, overdue, overlooked, overpaid, over-simplistic, over the top

P pandering to…, pass the buck (see bureaucrat), patronizing, pencil-pusher, pejorative, perverted, philistine, pinko, political, Pollyanna, pompous, prank, prey upon, pro-abortion, pro-crime, pro-something, promise

Q queer, question (as in 'that's not the question')

R racist, radical religious Reich, rarefied, red flag, repugnant, revolt, rhetoric, rift, road-block

S sad, Santa Claus, sarcastic, screw-up, secretive, selective memory or statistics, self-appointed, selfish, self-serving, senile, sensationalize, separatist, serial-murderer, serpentine, shameful, slash, slick, silent killer or majority, simmering, simplistic, smirk, snivelling, special interest, so-called, socialist, split, steal, stereotype, strike out, storm trooper, stumbling block, stupid, sugar-coated, sympathizer

T tardy, tart-up, tax dodge, tear a strip off, terrible, time-consuming, those people, torn down, tragedy, traitor, treacherous, treason, trivialize, turn back on, turn-coat

U ugly, undermined, unable, unbelievable, underground, uninformed, unpredictable, unparliamentary, unprepared, unrepentant, untrustworthy, unworthy

V vote buying

W wander in the wilderness, waste, weasel out of, weird, white elephant, wildly optimistic, winging it, witch doctor, wormed

X xenophobic

Y you people

Z zealot.

Now, these words all have their purposes. Reporters love them. Many are attack words. Some may be quite acceptable, in certain contexts or uses. If it's your plan to use them, fine, but be careful when they're contained in the reporter's question. The reporter is telling you something with the wording of the question. The questions may reveal the reporter's spin.

How would you answer them if you were attacked? Ask yourself a question using one of these words and then replace the word with one that you choose. The replacement word is the defining word. You're now a spin doctor.

If you use one of these glaring words in an answer, it will likely become a quote. Even and especially with the word "not" in front. You cannot ban the "not" word from your vocabulary. It's essential in many situations. You must, however, use it only when it's intentionally planned in advance.

Planting the pillar foundations

Decide on one word or phrase that will complete these statements and now you have your first pillar statement for your media line. If you need more than one word, your limit, at this stage, is three words.

Some of them are a bit shop-worn and overused, but we hear them every day on the news. Pick one or two, or develop a couple for your own use. Use them consciously and purposefully.

Issue-defining bridging phrases:

"The real issue is…"

"The concern we need to be addressing first is…"

"Our goal is to…"

"The fact is…"

"The situation here can best be described as…"

"What we're really talking about is…"

"What's happening is…"

"What's happened/is going to happen is…"

"This is a ________ issue, plain and simple."

"It feels like…"

"The result of this will be…"

"Our first priority is to…"

"Read my lips, it's a…"

"We have a responsibility to…"

"The real problem is…"

"We're talking about … here."

"There are really three issues here…"

"The fact of the matter is, it's about…"

"Our task/job/role is to…"

"We're here to help the police by..."

"The core of the matter is, we need..."

"Let's not forget what's happening to..."

"Our main priority/goal/objective/commitment/ concern is..."

"The point is, we're talking about..."

"We share their concern, that's why our goal is to..."

"Let's be perfectly clear, we're working to improve..."

"What we're talking about is..."

"At the end of the day..." (I hate this one.)

"What's at stake here is..."

"Let's look at what's been achieved. We have a plan/proposal/idea that will prevent/improve/ develop ..."

These are your first sound-bites. They're the central means to define the issues in the questions and take control of the agenda. Their purpose is to protect you and take control of the communications encounter while avoiding the words of others.

The use of the defining phrase sends out a strong signal to the reporter that you're in charge, and reporters are well aware of what you're doing.

Here are some variations and examples of issue-defining pillars:

"Photo radar saves lives."

"Operating costs are increasing, salaries have gone up three percent and distribution costs have tripled. That's why changes are necessary."

"We're constantly looking at ways to improve customer service. There are new programs in accounting, improved computer equipment and better training systems in place to meet the growing demand."

"Safety is our main concern."

"What we're talking about here is a matter of basic human justice."

"Truck safety starts in the classroom."

"Illegal drugs are everyone's business."

"We cover every aspect of the case in every investigation."

"The truth is obviously missing from the Minister's agenda."

"This is an issue of equal rights for all people."

"Product tampering is an industry-wide problem. That's why…"

"This accident was preventable. It resulted from human error and equipment error."

"The price increase is due to market forces in North America, increasing consumer demand and the Asian economic crisis."

"Our role is to administer the program."

"Our job is to inspect trucks."

"Ground transportation service has been improved as a result of this new contract."

"The increased fees result from the government's commitment to cost-recovery, so that those who use the ferry service will directly pay for it, rather than the general taxpayer."

"This situation could have been prevented — that's why we're examining every aspect of the accident, working with the regulatory agency and reviewing our training programs to determine all the contributing factors."

Now, read through these samples again, to see how they might apply to your issue. Re-write some of them and make them your own. These are your defining pillars — defining the issue, your concerns, your role in the issue, your main actions, the main factors or circumstances and much more.

Here's another way to get started

Start your media line by asking yourself the most direct questions and answering very briefly and in point-form.

Use one to three words to answer these open-ended questions:

What is your main concern?

What's happening, happened, will happen, might happen?

What are the most important contributing factors here?

How would you describe your actions/the situation/the public impact?

...your position?

...your role?

...your viewpoint?

For whom are you doing this?

With whom are you doing this?

Why?

How?

When?

Where?

These answers now cover a lot of ground —

- ☞ they define the issue
- ☞ they can define your role in the issue
- ☞ they can deflect the issue elsewhere
- ☞ they can buy time
- ☞ they can reflect on your professionalis,
- ☞ they can help lead you into the rest of your media line
- ☞ they can be your one and only quoted statement

Case Study: Caring about business — Escalator safety in shopping center

You're the manager of Westhaven Galleria Shopping Center. At noon, a teenager gets a foot trapped in an escalator and is rushed to hospital screaming "I'll sue your _____ off." The youngster is related to a city councillor.

A teenager hurt on an escalator can generate issues of public safety, building security or maintenance. Plus, related issues of awarding of maintenance contracts or cutbacks in operating budgets. Or the issue may be one of providing safety guards along the edges of escalator steps to prevent accidents by people catching their feet in the corners.

Yet this escalator incident could be due to youngsters running, pushing and shoving each other on the escalator while under the influence of mind-altering substances.

The real opportunity issue here may be how teenagers conduct themselves in your shopping mall and the need for teens to exercise safety practices. Or the issue may be re-defined as substance abuse in the community, if it is known, documented and provable that some young people are involved with using, selling and abusing drugs.

Your shopping mall management might want to urge teens to conduct themselves more appropriately, or you may decide to work with youth groups to offer after-school programs in a community center.

Your strategy may be to turn the issue into one of teen conduct rather than escalator maintenance. *This, of course, is the basis of a great deal of public relations practice — to change the definition of the issue in order to change the behavior or opinions of the public.*

If, however, there is major evidence that accidents can easily be prevented by installing new technologies to the escalator steps, we strongly suggest that you announce that they will be installed as soon as possible — and do so.

It's insufficient to say you care about safety without doing something about it.

The story can be a one-day wonder that quickly passes into history. Or it can drag out over five days, prompting the media to examine other escalator incidents and result in sweeping changes in inspections, maintenance and alterations. And if that is necessary, you'll want to be a leader, on the right side of the issue.

Speculations

Are you trying to fix the problem? Or are you fixing it? There's a big difference. Mark Twain once said that the difference between the right word and the nearly right word is like the difference between lightning and the lightning bug.

Can you guarantee that you'll fix it? Or are you helping or working to fix it? There's a difference.

Speculative questions can be among the most dangerous. Yet many media lines will be designed to predict how events may occur or to forecast future activity. These may be based on "expert claims" which your critics might refer to as "wild speculation." Is there a difference?

As a neighborhood spokesperson, your goal may be to predict what will happen to your neighborhood if a new factory is built nearby. But **as a city environmental officer** monitoring the development, you might not want to speculate on the environmental impact of the plant until experts see how it operates. **As a company spokesperson**, you will want to stress that air-monitoring equipment is designed to be effective. None of you can make guarantees.

As a student leader, you may want to make claims about what student loan cutbacks might do to the education system in the future. **As a bank spokesperson**, you won't want to speculate on what banks will do to recover overdue accounts for student loans. **As a university spokesperson**, you'll want to assure that alumnists keep donating to your growth fund.

As an opposition politician, you may want to assure the voters that a risky initiative will fail, yet **as a government spokesperson**, you might want to sit on the fence and watch how the wind is blowing before speculating.

A wildly speculative question for any of us is a career-ending opportunity, or a career-enhancing opportunity. The secret is in the answer we use.

Case Study: Caring about your professionalism — Highway tolls

God bless engineers. We've trained hundreds of professional engineers and our most favorite trap question is a variation of "Is it possible…?" Perhaps because of their training, personality types and/or their background, they look at this speculative question in a very narrow manner.

Here's a typical first question from the reporter from hell: "Is it possible that we could see highway tolls on Highway 6 through Westhaven?"

For some reason, over half the engineers answer with something like: "Well, I suppose that anything's possible, but…"

Now if you didn't actually hear the voice inflection or see the body language when this answer was quoted in a print article, the reader will only have the written word to rely upon for interpretation. Is the emphasis on the "suppose" or on the "possible."

You must assume that anything you say after the word "but" will be completely ignored and the front-page story tomorrow will be headlined: "Highway Tolls Possible: Tourist Board Outraged." I often avoid the use of the word "but" and replace it with the word "and."

Does this interview situation happen? If it didn't happen, you wouldn't be reading this book. It happens every day and problems can be prevented.

Have you been misquoted? Not that you'll be able to do very much about it — the damage will have been done.

Is it the reporter's fault? Chances are, you could have said it better. Don't blame the media for your communications deficiencies.

More importantly, as the project construction co-ordinator, you should have focused on your role and involvement in the project.

You should have said: "I'm not aware of any decision to change the present policies regarding tolls on state highways. Our main concern is the safety of the highway between… That's why we're building additional passing lanes, improving illumination at intersections and widening the shoulders along curves."

Other examples of speculative questions:

What if…?

Could it be that…?

How will others react?

What will others say?

What will others do if…?

Should you not have…?

Won't taxes increase if you…?

How many people have to die before…?

What will happen to… if…?

Might there be…?

Remember, your answers to all of these questions should be in your media line. Or the answer may simply be "I don't know. (Pause two seconds.) What we're working on is…"

Keep in mind that **speculative situations may be part of your strategy**. A neighborhood group may take the position that if a new highway is built, certain things will happen to real estate values.

As a neighborhood spokesperson, it may be your strategy to predict devalued house prices, unsafe pedestrian situations near schools, or truck tires flying through your living rooms.

Your credibility will depend on your believability, backed by facts, research and your experience. Expect the reporter to seek out collaborative or opposing views whenever an issue arises. Expect the reporter to be skeptical.

Your defining statements may be ones that serve to control speculation. In this case the issue may be the timing and decision making of an action or the process required to take a certain action where there is a great deal of controversy. If your expertise is in the process, focus on the process. If your expertise is engineering, focus on engineering.

Buying time

Your goal may be to buy time while the matter is being studied, so as not to be seen to be jumping to conclusions.

You'll likely want to draw attention to the thoroughness, expertise and professionalism of the investigation or study.

Or the need for the public's co-operation in solving a long-unsolved crime.

Or the amount of time it takes for certain information to become available from other sources outside your organization's.

Or to show that you're waiting for the actions of others before you act.

You may choose to handle the speculation with some of the following examples of super-controlling statements that buy time or deflect the issue. Some of these statements are exceptions to the "not" answers.

Super-controlling pillars:

"There are no plans to..."

"I'm not aware of any plans to..."

"No announcement has been made on..."

"There is no evidence of..."

"The investigation is continuing. Anyone who has information on this matter is urged to call Crime Stoppers at..."

"No final decision has been made regarding..."

"We are continuing to..."

"Labor negotiations are continuing. Our goal is to reach a settlement that's fair to our employees, fair to our shareholders and fair to our customers."

"The accident remains under investigation by the Westhaven Police Department."

"We are anxiously awaiting the outcome of forensic studies to provide us with additional information."

"All commercial vehicles traveling on state roads are subject to an inspection by a transportation enforcement officer. That inspection can include brakes, lights, security of load, manifest, driver's log..."

Chapter Eight

Never say "No comment"

Your job is to inform, describe, explain

There may be a number of situations where you or your organization do not wish to participate in media coverage. Your goal may be to stay out of the issue. You may not wish to comment on certain aspects of the issue. And for the vast majority of my students, your job is not to comment on issues anyway, your job is to inform, describe and explain.

If you're the mayor's executive assistant, do you want to comment on her latest controversy? Or do you want to explain why the city "is concerned" about the issue of homelessness? As a government spokesperson, do you want to comment on what politicians are saying, knowing that some day they might be your boss, knowing that they monitor news coverage very carefully?

Not commenting is one thing. Saying "No comment" makes you look guilty.

There are lots of reasons never to say "No comment"

- The phrase often serves to confirm questions like: "Is there a cover-up? Is the mayor an idiot? Will taxes increase?"

- You create an information vacuum where others will quickly move in to speak.

- You will make the issue worse.

- You're saying "No comment" directly to the reader, listener or viewer, suggesting that you don't care enough about them to deem to speak to them.

- Reporters love the phrase — it's short and concise and speaks volumes about your apparent level of public-spirited professionalism. It proves that the reporter has made an attempt at balancing the story.

- There are probably lots of other things that you can and should say instead — and they're all in your media line. (Or they should be.)

Often we see the phrase "No comment" when, in fact, the spokesperson intends to say "no answer", or "no response." Or when they really mean that they have "no information — yet."

Police officers have strict limitations about discussing evidence on a case in progress or before the courts, yet others involved in the case may have no restrictions on speaking. Police need to constantly remind reporters that they're not permitted to disclose evidence. Yet, if asked whether the mayor is a suspect in a criminal case, saying "No Comment" is the equivalent of saying yes. If that is your goal, then, and only in such situations is "No Comment" appropriate. There are a few exceptions to almost every rule.

Case Study: Caring about immigration — Deporting the maid

I used to work for Canada's immigration department. We often had to deal with the issue of deporting someone, and sometimes that person went to the news media with a tale of hardship. We could say nothing about individual cases, due to privacy laws. We were often made out to be the villains, sending folks back to poverty and sure death.

Sometimes the real reasons for the deportation (such as fraud or a criminal record) had not been revealed to the news media. The media portrayed the deportee as a saint.

What we learned to do was to advise the reporter that the reasons for the deportation were contained in a letter of a certain date and that the deportee had been provided with the letter.

If the reporter wanted the whole story, the letter explained the reasons for the deportation. The reporter was urged to obtain a copy of the letter from the deportee.

This practice reduced the number of such deportee stories and improved media relations.

"No comment" alternatives — When the first word is the last

Here are some statements that are far more appropriate than "No comment." They clearly indicate why you can't answer the question, or when you will be able to speak about a matter. These statements will not likely be used as quotes, but you must write them in a way that is open and clear, in case you are quoted saying them.

When you say you can't comment on something, there's a perception that you're withholding information from the public because there's something that you don't want people to know.

If there's a reason you don't want information released, and you're not required by law to release it, then you need to explain the reasons for the policy, instead of saying "No comment." You want to indicate why there are others better able to answer the question than you can. Be prepared to repeat these statements several times in an interview.

> **"I don't know the answer to that specific question at this time. I can put you in touch with someone who can help you. Or, I will research the information for you and get back to you. When's your deadline?"**

> **"I'm sorry, but the privacy act and regulations prevent/prohibit me from speaking about that specific matter. I'm sure, as a reporter, you're well aware of the privacy laws which are in effect in this instance. I'm sure your newsroom is just as concerned about privacy issues as Westhaven Hospital is."**

> **"We're not authorized to discuss any particular case or correspondence before the board, however on the matter of sexual misconduct charges, it's the board's policy to review each case beginning with ..."**

> **"We are not authorized to release any information on that question, because the release of that information might diminish our security efforts. We can tell you, however, that security is our number one priority and we're constantly working to improve security measures, for instance..."**

> **"Westhaven Shopping Mall does not provide any details on bomb threats for the same reason that newsrooms do not release information on security threats that they receive — because such release often leads to more bomb threats, many of which turn out to be hoaxes."**

> **"There is no evidence to support the allegation." (Or, "There's no evidence to suggest that... is connected in any way with... The investigation continues to focus on all matters relating to...")**

"The judge has issued a publication ban on the case which prevents me from speaking about matters before the court. A number of issues surrounding this case will be dealt with in the courtroom over the course of this enquiry/case/trial. We will be stating our position at that time."

"I have been advised by counsel not to speak to the news media at this time, because the matter is before the courts. I know that the truth will come out and I look forward to that day. Thank you for your interest in this matter and for your consideration in protecting me and my family from undue harassment."

"You've raised some of the same questions we're asking. We're anxious to know, for instance, ..."

"Under the terms of the agreement/contract/judge's order, I am prohibited/prevented from discussing any aspect of this matter."

"As a private company, we do not release certain financial data which may be detrimental to our competitive position."

"I have been instructed to refer all media enquiries to our public affairs spokesperson, Ian Taylor. His telephone number is (416) 466-1560."

"There are a number of experts on that topic who are better able to answer that question. Our role in this situation is a financial/advisory/consultative one. We're..."

"Those questions are best directed to the police department who are investigating this situation."

A reputation for being ready for anything

To decide how to develop your defining statements, determine what questions are probably or possibly going to be asked. If you believe that it is at all possible that the reporter might ask you about a certain topic, issue or sub-issue, decide what your answer will be. "If asked about this, stress this…"

Pose questions to yourself in the most negative way. Fill the questions with glaring, inflammatory and negative words. Create worst-case questions, so that if the reporter asks one, you'll be ready with an answer.

Make the questions nasty. Make the questions angry, abusive, accusatory. Or, like some reporters we know, ask the horror questions gently; the technique is often more effective than some hostile interrogation techniques.

I use all of these techniques in our training courses. That's why your author has been called "the reporter from hell." Don't hold back in preparing for the most negatively-phrased, inflammatory or glaring questions — you just might be asked one in an interview and you must be prepared, no matter what the subject of the interview. "Will you profit personally from this development?"

Here's where your personality type comes to the fore. When asked an ambush question, will your eyes or body language give away the real answer? Will you get defensive? Will you respond with anger?

> **Anger is the enemy of reason.**
>
> — *Anon.*

Learn to create your own examples of "reporter from hell" questions. In each case the question could have been worded: "Can you please bring us up to date on what you are doing about…" Or, "Could you describe the situation?" That's the wording that your mind must hear when answering the horror question.

Instead of opportunity words, the reporter is purposely using glaring or inflammatory words, false premises, leading accusations, loaded assumptions or stretching the facts. The reporter is using the questions to attempt to define the issue — and spin is your job too.

The reporter controls the editing process. The reporter can remove the question from the news story and you can be left talking about items in the questions that have nothing to do with your message.

Does this happen very often? I've been interviewed thousands of times and it's happened enough to me to want to warn you.

You start each of your answers to these horror questions with a bridging phrase like these:

"Let me tell you what we're doing."

"Here's what we have to do first."

"Let me explain what's happening."

"Let me start by bringing you up to date on this situation."

"We're going to start with..."

"Our first priority is to..."

Favorite questions from the "reporter from hell"

"Are you able to sleep at night knowing that people are dying while you mismanage this project?"

"Would you describe this as a disaster, a screw-up, a mess, or a crisis?"

"Is there a cover-up?"

"What's your personal opinion on this?"

"Why did you wait so long to act on this issue?"

"Why are you playing fast and loose with the facts here?"

"Is this a glossy public relations exercise to cover up the truth?"

"Is the mayor playing politics with this issue?"

"Are you involved in this for profit or personal gain?"

"Whose fault is this?"

"Are you in bed with special interest groups like…?"

"Do you agree with the unions who say this is a disaster?"

"How do you think this organization will react?"

"How many people have to die before you fix this packaging?"

"Will it take another death of an innocent, precious child before your company stops this campaign of incompetence?"

"Is your company so concerned with profit that you're prepared to sacrifice citizens with killer widgets?"

"How do you manage to look at yourself in the mirror knowing that truck tires are killing state motorists?"

"Would this situation have happened if politicians had met their responsibilities, instead of kissing up to their friends in the special interest groups like the chamber of commerce?"

"You've been covering up this situation, haven't you?"

"Don't you think people have a right to safe highways which are free from irresponsible trucking companies?"

"How would you feel if it was your daughter killed or maimed in an accident — wouldn't you be doing more?"

"What would you say to the innocent victims who are lying in their graves with six feet of dirt on their faces?"

"Would you describe the new facility as a white elephant?"

"As a professional ______, are you embarrassed at the results of this project?"

"Do you care more about profit than you do about people?"

"Was this decision politically-motivated?"

"Why have you ignored the concerns of…?"

"How much longer can we tolerate this buck-passing and finger pointing?"

Ask yourself the tough questions first

In each question above, you must learn to answer starting with a bridging phrase like, "Let me explain what we're doing and why we're doing it." "Our main concern is... That's why we're working with..."

Apply some of these horror questions to your issue. Re-define the issue. You'll end up with dozens of answers for your media line and you'll be better equipped to meet the reporter from hell.

Now, we're only slightly exaggerating these questions here. While it is unlikely that most reporters would ask questions like that, you can and must be ready for them. These questions could arise in any number of situations — a stockholders meeting, a briefing to an organization or at a coroner's inquest. You could get them at a public meeting or at a legislative enquiry. You might get them in a letter or a customer service encounter, or a hostile telephone call.

Remember, it's not the question that will get you into trouble, it's the answer.

Every nasty question can be interpreted in such a way that it provides you with an opportunity to communicate on **your terms**, with **your message**, building upon **your strategy**. Your spin.

Each of the nasty questions above contains hot-button, glaring and inflammatory words, just like the ones in the previous list of denial words. The questions attempt to define the spin.

Each of the above questions can be answered the same way — with a core statement from your media line that answers why, how, when, where or who. Go to message.

Ask yourself open-style questions

- ☞ What are you doing about this issue or situation?
- ☞ What do you want others to do?
- ☞ Describe the situation.
- ☞ What's happening?
- ☞ Explain your position.
- ☞ Can you be a bit more specific?
- ☞ Who are you working with?
- ☞ What are the main factors in this issue?
- ☞ What are some of the key statistics or facts?
- ☞ How are you dealing with this issue?
- ☞ Can you give me an example?
- ☞ What is your main concern/goal/objective?
- ☞ What is your personal involvement or motivation, if any?
- ☞ What will happen if…?
- ☞ What's the team and the plan?
- ☞ What needs to be done first? Second? Next?
- ☞ How can the public help?

And the most basic questions such as:

- ☞ Why?
- ☞ Why not?
- ☞ How?
- ☞ How much?
- ☞ When?
- ☞ How soon?
- ☞ Where?
- ☞ Who?

When you've prepared short, concise, factual answers to these types of questions, your media line will be well on the way to handling the toughest questions. Almost all the tough questions will deal with the same topics or issues that the above questions raise. They're just asked differently.

For most interviews, you'll need at least three pillar statements to deal with the main issue you've identified. You may need at least one more pillar statement for each sub-issue. The more completely you identify all the issues and sub-issues, the more extensive your media line will be.

Remember, your pillars are designed to meet the media's requirement for concise, clear and defining answers and quotes.

In several of our examples, we've provided more than one pillar. These give the reporter a choice of sound-bites. In fact, your first answer should provide the reporter with several quotable quotes.

This meets the media's major requirement for a 5.2 second sound-bite. The bottom line is, if you want to achieve a professional-sounding, public-spirited quote of 5.2 seconds, develop your own and use it. And be prepared to repeat it, to stay on message.

Your first pillar is your issue-defining statement.

This statement defines or redefines the issue in **your** terms, sets the boundaries for the interview and serves to control the agenda.

This may be a statement which you've been using for some time, or it may change to meet the changing nature of the issue or the existing public opinion environment.

Your second pillar may be your role-defining statement.

Find one to three words that describe your job in the issue. The larger the issue, the more important it is to establish your role in it to protect you from questions outside your area of expertise, authority and function.

Brevity and conciseness are the keys to successful pillars. You want to say as much as possible in as few words as possible.

Is it on your business card?

A recent trend in some organizations is to print their mission or goal statement on the back of business cards and we've seen several that average under 10 words. Shouldn't you make an effort to use this statement in an interview? You should use it if it meets these criteria:

- **The boss would be pleased if you were quoted saying it in an interview.**
- **Your organization has gone to great lengths (and costs) to develop it.**
- **It reflects on your and your organization's public-spirited professionalism.**
- **It is true.**
- **Most importantly: It is written in plain, conversational language that you're comfortable saying.**

Unfortunately, most of the mission statements I've seen look like they were written by a committee without writing skills. They often make unbelievable claims about being "the best." Is it credible? Unless the statement is clear, concise and specific, avoid using it. Or develop one that meets these criteria.

The alternative to a highly formalized goal statement is to deliver concise pillars that state your main concern, your main priority, an explanation of the situation, your next step in the plan, an identification of what your organization is working to achieve or a statement of purpose. In plain, every-day language.

Define your role in the issue

If you are a technical expert, rather than a full-time public affairs employee, you will need a pillar which states your job function or role in the issue. If you're the food services director at Westhaven Hospital and the reporter starts getting into non-food issues, you'll want to re-direct the interview by reminding the reporter of your role.

There are situations where you can use the words "I" or "my" and then switch to "we." Examples:

> **"My role is to administer the food services division. Our goal is to find cost-savings through areas such as…"**
>
> **"The role of Widgets Canada has changed as a result of new directions announced in the last budget. My office's role has changed from a funding agency to a monitoring one. We now work with … to assure… so that…"**
>
> **"I will be involved in the financial planning aspects of the new hospital. We'll be studying areas such as…"**

Samples of role-defining pillars

We've trained a number of highway safety experts and we invariably (and purposely) ask interview questions that fall way outside their areas of expertise and responsibility, such as questions about photo radar, like:

> **"Was it a good idea to kill photo radar?"**
>
> **"Do you think the government moved too quickly?"**

> **"Did you cover up information that showed that speeding was declining?"**
>
> **"Is it now a free-for-all once again on our highways?"**
>
> **"Would we need photo radar if we had enough police officers on our highways?"**
>
> **"Did it make common sense to sacrifice safety for political expediency?"**

The highway safety inspectors, for instance, are urged to use these types of answers promptly, when the first unwelcome question is asked:

> **"The job of an inspector is to carry out random inspection checks of commercial vehicles. In fact, every commercial vehicle traveling on state roads is subject to an inspection that can include items such as brakes, lights, security of load, manifest, dangerous goods, weight, height…" (Pillars.)**
>
> **"Let me give you an example of some of the problems we found at a recent truck inspection blitz near Riverdale. Over 55% of the trucks we inspected were found to have…" (Sparkler.)**
>
> **"All commercial vehicles are subject to an inspection, and if they're not safe, they won't move. And if they don't move, they don't make money. And if they don't make money, maybe they'll change their minds about safety. The safety of motorists is our number one concern. That's why…" (Pillars plus support.)**

Pillars that frame or re-frame the question

It's said that election campaigns often center on a political party's ability to frame the question (really, the issue or concern) in the voter's mind. Questions like:

Who can best meet the needs in the future?

How many more hospitals and schools must close?

What will be our needs for the future?

What kind of future do we want for our children?

Are you better off than you were ten years ago?

Framing questions are helpful in leading off an interview when the reporter's question is unwelcome. When you frame your own question, you obviously have the answer ready.

The framing is essential when the reporter throws an ambush question at you early in the interview. Then the **bridging** phrase is, "The question we need to be asking ourselves first is…?" Then, answer **that** question.

Chapter Nine

Concise, direct and interesting

> **As the TV character, Ricky Ricardo used to say, "You got a lot of splainin' to do, Lucy."**

Can you start with two-point or paradox pillars?

"On the one hand, prices are increasing. On the other hand, so is level of service. That's why..."

"You've raised two separate issues in your question. First there's the matter of..."

"In making our recommendations we must balance the need for... with our concern for... That's why we need to study..."

"There are two sides to this issue. On the one side, there's concern for..."

"Our policy must not only take into account the issue of..., but must also address our concern for..."

"We've got good news and bad news. The bad news is that..."

Paradox pillars are helpful in buying time on an issue. They illustrate why action can't be taken immediately. They foster public discussion and they force your supporters and opponents out of the woodwork.

Issues are seldom two-sided, however. Often it's the middle-ground that is finally reached, a consensus. Things are neither black nor white, but some shade of grey. That's one reason three-point statements are so important.

Taylor's top ten reasons to use pillars with three's:

- **They're relatively easy to remember — for you and your audience.**
- **They serve to avoid over-simplification or "single-factor analysis" by the media.**
- **They display your professionalism, knowledge and credibility.**
- **They allow for structure and follow-up context.**
- **They convey more information.**
- **They require you to develop clarity which assists the reporter in developing a story outline.**
- **They allow for greater understanding of the issue.**
- **They provide a sense of symmetry, of shape, of balance.**
- **They flow.**
- **They cover a lot of ground.**

Remember that lists of ten (and also lists of seven and twelve) work well in print or when you can display them graphically like on the David Letterman TV show. They're also great for brochures, web sites, articles, speeches, pamphlets or publications. But when you're speaking, three's work best.

A reputation for teamwork

We spoke earlier about the communications opportunities that exist when we remember to describe "the work, the team and the plan."

Can you explain the team's three major components?

Now, can you describe the three main parts or steps of the plan?

Can you outline three main benefits of your proposal?

Are there three unique features to your plan?

Are there three things people can do to be part of the solution?

Can you do it all in tightly-written, conversational sentences under a certain number of words? You've now got pillar statements for a media line. You're starting to spin.

Strategic tips for using three's

Avoid referring to the number three in your statement, in case you forget one point. Each of the three items will naturally be important to you, although the reporter may not agree with your priorities and may want to focus on one of the items. You'll need to refer to three only if your answer is long and complex. In short, focused statements, the listener will keep up with you.

When you make three points in a TV interview, try **pointing on your fingers** — it's an effective visual tool, but hold your hands just below your chin, so that they're "high and tight."

I like to **place the most important issue third**. It stays with the recipient longer and allows you to build on it. Decide what feels best for you on an issue-by-issue basis, but make the conscious effort to order the items strategically. The alternatives are to arrange your three's chronologically or according to priorities.

Factors and factoids

Here's how you might develop three-point pillars and lots of examples. The reporter can use all three points, or, in some cases, you've provided three stand-alone points that the reporter can use to frame a three-part news story. That means a 45-word answer can create about three 15-word responses, give or take a few words. That's three quotes — more than enough for many interviews.

Identify three main factors:

"Accidents continue to mount on the new highway. The costs of responding to accidents is increasing. Most of these accidents are preventable. That's why…"

"Food costs are up 12.4 per cent. Labor costs are up 9.2 per cent. Menu prices are only up four per cent. We think that's fair."

"Our new safety policy involves education, compliance and enforcement. Our goal is to save lives, reduce the personal tragedy and contribute to lower insurance costs. To do that, it starts with drivers. They must obey the speed limits, wear their seat belts and stop drinking and driving."

A variation on two sets of three's — three facts, three framing questions:

"Health care has been underfunded, people are dying while waiting for surgery, and hospitals are closing in many areas. Is this what our government promised? Is this fair to Ontario's poor, sick and elderly? Is this common sense?"

Some self-serving free publicity with sets of three's:

"Our previous training programs were too expensive, too generic to meet specific management needs, and too slow to respond to issues we face every day."

"Never Say "NO COMMENT" Incorporated offered us real value for our training dollar. They matched the training to meet the issues we face every day, and their consultants were available quickly when we needed them."

"Now, when a reporter calls, we know what to ask the reporter, how to develop a media line, and how to deliver it quickly."

Or four sets of threes, separately or combined:

"Customers are seeking value, service and quality."

"They want products that are easy to use, safe and environmentally friendly."

"They want instructions that are easy to read, easy to understand and don't require them to call on their teenagers to help."

"That's why we've introduced the new widget, developed new instructions on assembly and set up a toll-free customer service line at 1-800-WIDGETT."

Name three benefits of the new program/product or service:

"This new widget will save time, reduce costs and protect the environment."

Or:

"Today's clothing consumer wants more than just a pretty dress."

"She wants fashion, style and value for her clothing dollar."

"She wants her clothes to serve her at work, for private social events and for those other special events that are part of her new professionalism."

"That's why Westhaven Fashions has introduced a new fashion line called Today's Woman. The line features…"

Whoever seeks all the credit can end up with all the blame.

"Westhaven Police will be working closely with neighborhood youth groups, local businesses and area high schools to develop a plan to deal with crime among young people. Our goals will be to…"

"In order to improve our travel policy, the finance office brought together people from operations, human resources and senior management."

"We developed changes that met the needs of frequent travelers. They're fair to everyone. And they achieved cost savings of over 12.5 per cent."

"The new travel policy means faster trip approvals, reduced waiting time for reimbursement of costs and an expanded list of accommodation options. Here are some different examples…" (Pillars *and* sparklers, here.)

Provide three steps people can take (the plan) to do something:

"In order to make your business accessible for the physically-challenged, start with the doorways, the washrooms and the floor surfaces."

"Train your staff to focus on their listening skills, their observation skills and their customer service skills."

"Watch your business improve, your profits increase and your goodwill skyrocket."

Outline a chronological sequence in three steps:

"Once you've purchased your new widget, assemble it according to the enclosed instructions. Wait until the glue has dried before using it on a clean, flat and sanded surface. Cover the surface a second time and you've got a counter-top like new."

A three-point safety lecture format:

"Before heading out onto the waterways this summer, the Coast Guard recommends that your vessel be water-ready."

"Have your boat's engine safety-checked, replace all outdated flares or safety equipment and check the latest navigation charts for any changes in waterways in your area."

An accident investigation model:

"Before we know what caused the accident, we want to investigate the mechanical, electrical and structural components of the vehicle."

"We'll talk to the drivers, the passengers and any witnesses."

"We'll study the road surfaces, the traffic control systems and the visibility factors."

"We'll be working with the medical examiner's office, the insurance investigators and government safety experts."

Provide three options:

This can sometimes be difficult, but typically, there is often the 'do-nothing' option with its consequences, the radical, high-cost option, and the preferred or recommended option, in that order. In each case, you might also develop three consequences or results of each option.

Power pillars

Thinking about three's, writing in three's and speaking in three's forces you to be concise, orderly and interesting.

☞ Three's help the reporter and you to be concise.

☞ Three's are an essential part of many news stories.

☞ Three's help the spokesperson and the reporter, to structure messages.

If you don't believe it, check several newspaper articles to observe how often reporters write in three's and how often spokespersons who speak in three's have their quotes used.

The *MediaSpeak* Power Pillar formula involves defining the issue or concern, followed by a three-point statement. "Our main concern is this. That's why we're doing this, this and this."

The click

I've been interviewed by reporters on thousands of occasions — by phone and in person. I've also talked to other spokespersons about a phenomenon I call "clicking," with apologies to Faith Popcorn. (Read her books.)

It happens like this: You're answering a question with a multi-point answer that flows very nicely. Your mind suddenly clicks that what you're saying is what is going to be used in the final news story. In over half the situations, it happens.

Maybe something clicks for the reporter, too. Maybe you've just been lucky, but in the middle of saying it, you feel the click.

A reputation for order

☞ Explain what usually happens, what sometimes happens and what happens rarely.

☞ Say what most people do, what some people do and what a few people do.

☞ Indicate what it normally costs, but that it can cost as much as this or as little as this.

☞ Tell when it is likely to happen, and that it may happen as soon as… or as late as…

☞ Your job usually entails this, but it sometimes includes this and once in a while this…

Chapter Ten

Supports — Know when to hold 'em & when to fold 'em

Highlights

What are Supports?

- ☞ Two types: Short literary devices and one-liners that support your pillars
- ☞ Statements which support you in the interview process

One-Liners:

- ☞ These generally average under 15 words, sometimes one word
- ☞ Short, plain language
- ☞ Metaphors, analogies, similes, clichés, figures of speech, idioms, slogans, mixed metaphors, axioms, proverbs, euphemisms, famous short

quotes by others, clinchers, clever phrases, new applications of old words, trendy words

- ☞ Titles of reports or speeches, books or brochures
- ☞ Reflect our culture of work

Interview supports:

- ☞ Phrases that help you to deal with the tough questions, or to keep reporters in line, or to manage the interview relationship
- ☞ Bridges, baiting statements, repetition, re-asks, pauses

How do you use them?

- ☞ In support of your pillars, or can stand alone
- ☞ Do not over-use
- ☞ Anytime in the interview, including lead

Why do reporters like them?

- ☞ They sum up a lot in a nutshell
- ☞ They account for about half the sound-bites we see, over 75 per cent in sports news
- ☞ They meet the need for plain talk
- ☞ They can condense complicated issues into easy-to-understand idioms or phrases
- ☞ They're part of our lexicon
- ☞ They reflect your personality, emotion, knowledge
- ☞ They're comfort food
- ☞ Reporters often can't get enough of them in an interview

How to develop them:

- ☞ They're already all around you — in everyday language, in marketing talk, annual reports, language of work
- ☞ You will get comfortable using them
- ☞ Select a favorite quotation
- ☞ Answer questions like:
  - Are you part of the problem or part of the solution?
  - If this were a baseball/football game or horse race, where would we be?
  - What do we need to do to get off the ground?

Keep them short, sweet and to the point

Do not use too many one-liners, unless you're an athlete, in which case you can have dozens. Remember the movie *Bull Durham,* in which the senior baseball player teaches the rookie to use three or four clichés to employ in every media interview? These were media line statements, or "supports." And in the movie, as in real life, they generate headlines and quotes.

Supports often prop up pillars. They work best when they're delivered before or after a pillar statement, sometimes using the word "because."

Supports can also stand alone. They often serve as a catchy title to a speech or presentation, a report or book, or even a company name, like **Never Say "NO COMMENT" Incorporated.**

Literary devices like metaphors often result in headlines because they're so brief and catchy. Examples:

> **"Government pouring money down drain, critics charge"**

"Election enters final stretch, Governor tiring"

"Mayor shoots for Olympics"

"United Way running dry"

"Giving welfare recipients a hand-up and not a hand-out"

"Bureaucratic boondoggle leads to Transit overruns"

"Strike an 'act of faith,' union declares"

"The common sense revolution is nonsense, protesters charge"

Support statements meet the media's needs for clichés, figures of speech, metaphors, slogans, euphemisms, analogies, similes, platitudes, banalities and triteness. These are classic literary devices in all the traditional forms of writing and you seldom have to look far to find clever words.

The *Book of Proverbs* tells us "A word fitly spoken is like apples of gold in pictures of silver." And elsewhere, "a good name is rather to be chosen than great riches." Those phrases express how I feel about communications and reputation management and I often use them in my speeches.

Anyone who's studied Shakespeare, the Scriptures or other famous writings, will know the richness of language through the tools of literary devices.

The media has an insatiable appetite for the cliché. Just check the sports news and count the clichés. Watch how often horse racing terms are used during an election.

Reporters love clichés — even though they deny it

While some journalists criticize the use of clichés by public figures, some writers couldn't operate without them. Just look at this book.

Example: *"Internal Bleeding."* An article about the Canadian Red Cross by André Picard in *Saturday Night,* October , 1996, p. 31.

Here are a few edited excerpts, emphasis ours:

> "The Red Cross is no longer a **sacred cow; it's a moving target."**
>
> "...the blood program has become a **financial cancer**."
>
> "An **iron-fisted** administrator..."
>
> "The plea (for funding) **bore fruit."**
>
> "In such an arrangement lay the **seeds of disaster**."
>
> "...**half-baked schemes** to build fractionation plants..."
>
> "...whose experts **wore so many hats**..."
>
> "...it's big business, at **the cutting edge of science**..."
>
> "As the Red Cross marks its 100th anniversary in Canada, it is a perfect time for a change in direction, **a rebirth**."

It's unlikely that the Red Cross was pleased with the coverage, nor would any of these clichés likely be part of their official media line. And not a single one of them is attributed to anyone — they all result from journalistic licence, or the abuse thereof.

The reporter has chosen to use these worn-out phrases. But then, he's not the first or the last. Has the reporter moved from reporting to editorializing? Yes. This is an example of the old saying that the best time to kick someone is when they're down. The Canadian Red Cross is no longer in the blood collection business.

Anyone who watched the 1996 US Presidential debates will remember Bill Clinton's famous line to Bob Dole: "That dog won't hunt." Or Gary Hart being asked "Where's the beef?"

Or remember the senior citizen who told Prime Minister Mulroney if he touched pensions it would be "Goodbye, Charlie Brown."

There are some who object to the idea of sloganeering, but that hasn't halted the practice. My clients involved in safety would have their hands tied without safety slogans. (Metaphor.)

MediaSpeak from marketing language

In most communications and marketing programs, a major effort (cost) is put into developing catchy phrases and slogans, some of which are trademark terms that are recognized immediately:

"Something special in the air"

"Over umpteen billion served"

"We want to be your store"

"Where quality is job one"

Then there's the extensive work that goes into developing various slogans or marketing messages for specific campaigns:

"Attend the Metropolitan Community Church of Westhaven. We're more than just a pretty faith."

"Westhaven International Airport is growing to meet demand."

"Westhaven Chamber of Commerce — a partner in progress for nearly a century."

"Westhaven Airport is committed to safety, efficiency and economic opportunities."

"Safe driving. It starts with you."

"A tradition of service and a commitment to community."

Often, these marketing messages have cost your organization a lot of money. They're a key part of corporate advertising and other communication. Shouldn't you be using them in your media interview? What's the risk in using them? The reporter may not consider them newsworthy, but you'll never know unless you try. Some would suggest that you're only sloganeering, but Goebbels told Hitler that the right slogan on the right poster could stir a nation to greatness.

Your use of slogans in an interview will largely depend on your comfort level. If you're very uncomfortable using such statements, and if you have the power to do so — change them to ones that feel comfortable.

Another option is to make yourself comfortable with them by practicing them in simulated interviews. Your third option — swallow your personal opinion and use them anyway to display your apparent team-spirit. As the change consultants will tell you — "get used to it, get over it or get out of the way."

My Aunt Betty had quite the reputation

Any public speaker will have on the shelf, a book of famous quotes or sayings. One of the most respected is *Oxford Dictionary of Phrase, Saying and Quotation,* Oxford University Press. It quotes William Randolph Hearst as having said, in 1898, "You furnish the pictures and I'll furnish the war." Wag the dog, anyone???

We have a technique when we want to use a worn-out cliché or when we want to be seen to be distancing ourselves from the quote. We'll say something like:

> **As my Aunt Betty used to say, 'you get what you pay for.'**
>
> **As the old saying goes, 'time heals all things.'**
>
> **One of my university professors used to say, 'the best place to start is at the beginning.'**
>
> **As our chief designer says, 'this is as smooth as silk.'**
>
> **Please pardon the cliché…**

When you use a documented or famous quote, attribute it to the author whenever you know the name of the author. If you cannot remember the name of the author, say so.

One of my favorites, attributed to several men so I purposely paraphrase with the word **that**: "I remember reading somewhere **that** laws are like sausages. If you have any respect for either, you won't want to watch them being made."

US President John Kennedy stirred a nation when he said, "Ask not what your country can do for you. Ask what you can do for your country." Yet it was Oliver Wendell Holmes, Jr. who had said, "It is now the moment… to recall what our country has done for us, and to ask ourselves what we can do for our country in return."

Quotes can be an essential part of speech-writing. We can think of several high-priced speakers who have never been known to have an original thought. Yet, they demand big fees to package the messages of others and make them relevant to their audiences.

Many of my clients recoil when we talk about using clichés. And it's always tough to think up a few examples in a hurry. If you need to, consult a book of quotations and search for ones that best sum up your point or position. Chances are, it will speak volumes about how you feel. And when you've found one you're comfortable with using, use it. This is an important part of the Toastmasters learning method — there's a Quotemaster for every meeting.

Reporters often prefer support-type statements to pillars, especially as the story proceeds and your position or main concern is already known. Here, the one-liner becomes the message of the day. "I think the mayor's performance was worthy of …"

Malapropisms and old favorites

Yogi Berra mangled metaphors. Whether it's done cleverly or accidentally, the results can be terrific — and, of course, disastrous if delivered poorly. Some are expert at purposely mangling metaphors or creating malapropisms.

The secret is to choose them carefully for public consumption and think about them in advance — there are too many metaphors that may contain language which may offend certain members of society and will cause later problems for you.

Here are some variations of one-liners:

"Blaming the media for misquotes is the last refuge of a scoundrel."

"The road to hell is paved with good intentions and it runs past the Minister's door."

"The government cares more about Wall Street and Bay Street than it does about Main Street."

"Attacking a reporter is like kicking a skunk — even if you manage to get in a kick, you'll still end up stinking."

"There's no sense hiring a dog if you intend to do all the barking yourself."

"It's the answer, stupid."

"Reporters would rather eat oats that have not traveled through the horse."

A reputation for "humbility"

In planning your media lines, be careful in choosing adjectives which cannot be measured or which sound like opinions. When talking about yourself, don't use bragging words. Avoid "excellent, top-notch, terrific, fabulous, tremendous, very good." Without establishing credibility, the spokesperson is not always believable.

Have you been paid to say this?

Do your words leave you open for criticism or skepticism? Did you find them in some self-help guru's list of power words? Don't oversell. It may be appropriate to use such terms about others, but not when you're seen to be congratulating yourself or your organization.

Declaring yourself to be the best is almost always arguable. If you've won an award for being best, that's different. Use measurable words — "award-winning, best-selling, fastest-growing, largest, first, only." When talking about your staff, they're "dedicated, hard-working, well-trained." Your work should be demonstrably "thorough, detailed and comprehensive."

Or use flattering words as part of a goal statement, such as "committed to excellence," or "our goal is to be a responsible neighbor" or "we're constantly working to improve…"

Note to entertainment publicists: write a news release without using the word "fabulous." Please.

Euphemisms

Instead of layoffs, firings or the slashing of employees, how often do we hear the terms re-structuring, downsizing or enhancing efficiencies? A euphemism is a replacement for an offensive word or term with one that is less likely to create controversy.

While some terms are blatantly unbelievable or constantly overused and patently unacceptable, a spin doctor is often required to come up with ways to soften the bad news through the choice of words. Therefore,

- A problem becomes a challenge
- Complaints or crises become concerns
- A spin doctor calls herself a consultant in reputation management

- Mistakes become valuable lessons
- Failures are called learning experiences which will help to keep it from happening again
- Instead of abdicating your responsibilities, you're changing the way you do business
- A cover-up becomes a commitment to confidentiality
- Closing offices and laying off good people becomes improving efficiencies and enhancing customer service
- Fixing mistakes becomes improving system integrity

There's a book on this subject called *Doublespeak* by William Lutz, proud presenter of the doublespeak award. The book is published by Harper and Row, New York, 1989. Lutz talks of "how government, business, advertisers and others use language to deceive you."

I've always recommended **openness, candor and the shouldering of responsibilities in all media strategies**. I've told students that there's no way their media line is believable without proof.

The bottom line is the careful selection of your words, in plain language, delivered from the heart about what you're doing that shows that you care for people.

Analogies are like tools

The analogy is a valuable teaching and explaining tool. When you're trying to explain something that people may not understand, analogies allow you to compare it to something they do understand.

> "This new office security program works just like Neighborhood Watch..."
>
> "This new education policy will result in schools that look like..."
>
> "The governor's announcement went over like a ..."

"This new legislation will result in a policy that is like..."

"Owning a Widget 3DX is like driving a ..."

"Applying for the program is like ..."

"Applying for this insurance policy is as easy as one, two, three."

Comparisons are an important way of putting an issue into context. A story on legalized gambling by Donna Laframboise, in the *Globe and Mail*, February 21, 1998, pointed out that Manitobans spent more on gambling in 1995 than they did on food. A later metaphor: "It is for good reason that gambling has been called a tax on the foolish. For every Cinderella story about a laid-off nursing assistant who becomes a millionaire, there are legions of losers."

Control supports

There are a number of absolutely essential phrases and techniques for spokespersons to use in interviews, especially when the interview is live to air, and the listener hears the question, the answer, the next question. These can be in the studio for radio or television, on the telephone for radio, or at the scene of an event by remote broadcast. Some are live to tape for later play and may be subject to editing. Find out ahead of time how much editing the material will be subject to. Ask the reporter before the interview.

In a live broadcast interview of about five minutes, (like *Canada AM* or a US morning show) you will likely be introduced by name, followed immediately by the first question. Acknowledge the introduction and, as necessary, bridge to background or define the issue. Use the reporter's name in a live interview, but not in those that will be edited.

Q. *Our guest today is Ian Taylor, President of Never Say "NO COMMENT" Incorporated. Ian, are you teaching your students to lie?*
A. *Good afternoon, Barbara. We've taught over 10,000 students and in my book,* ***MediaSpeak****, I deal very directly with the matter of telling the truth. The book says, on page____, that...*

And at the end of the interview:

> **Q.** *Our guest today has been author, Ian Taylor… Thank you for speaking with us today and remember, Ian, never say "No comment".*
> **A.** *Thanks, Barbara, it's been my pleasure. (Even when it hasn't.)*

Bridges

You'll never succeed at tough interviews without using bridges. They're the oldest tool in the interview training business. Just avoid the old ones, like "Let me put this in perspective."

Bridging phrases are highly strategic — you're telling the reporter you're exercising control without becoming a victim of the wording in the questions. Obviously, you have something to say that reflects on your public-spirited professionalism.

> **Q.** *Did you bungle this clean-up?*
> **A.** *Let me explain what went wrong. We had problems with the weather, the equipment and the procedures. At the time, the weather was…*

> **Q.** *Is there a cover-up?*
> **A.** *Let me bring you up to date with the latest information.* Or,
> **A.** *Let me be very clear about how we've handled information so far. What we've done is this, this and this.*

> **Q.** *Did you act too quickly?*
> **A.** *Let's look at the timing of this event. (Bridge) This happened, this happened, then this happened.* Or,
> **A.** *Let's look at what was happening in 2003 when we started…(three things.)*

> **Q.** *Are you playing fast and loose with the facts?*
> **A.** *Here's what happened. According to the police report…* Or,
> **A.** *Let me explain our accounting procedures. (three highlights, three commitments, three concerns)*

> **Q.** *Is this a case of corporate greed?*
> **A.** *Let me be very clear about our goals. (Issue-defining bridge) Since 2002, we've been working with the community to do this, this and this.* Or,
> **A.** *Let me give you an example of some of the work we're doing with 14 community groups, ranging from the boy scouts to the hospital auxiliary. (Bridge to a sparkler)*

What you're doing here is using a support tool — a bridge, to deliver your pillars and sparklers. You're also offering spontaneous answers to tough questions. You're volunteering essential information early. You're talking in sound-bites.

Baiting supports

In a live interview, there's a useful technique that's almost guaranteed to get the reporter to ask a follow-up question. You stop after making an enticing statement, like the preceding sentence. Now, wait for the live question — "what is it?" It's a method called baiting, and I just did it. The technique only works well in unedited interviews like open line shows or in-studio encounters. In an edited interview, keep talking.

> There are several myths about the media interview. ***Stop. Question:*** What are they? ***Answer:*** They include…
>
> There's one question that many customers ask us every day. ***Stop.*** What is it?

Phrases to avoid in an interview

The following phrases are argumentative and create negative spin. Replace them with phrases like the ones here:

> "No comment." Instead, say, "Yes, we're concerned about…"
>
> "That's not the issue." Instead, say something like, "The real issue is…"

"That's not important." Instead, say, "What's most important right now is…"

"I'm not prepared to answer that." Instead, say, "I would be pleased to answer your questions when I have more information (or once the case is heard in court, or after the meeting). Right now, we're…"

"I can assure you." Honey, you can assure me that the sun will come up tomorrow and I won't believe you.

"To tell you the truth…" (Were you lying before?)

"To be honest with you…" (Would you be otherwise?)

"Those are your words, not mine." Instead, say, "Our position is…"

"I wouldn't use the word lemon." Instead say, "Yes, there were some problems with the satellite."

"My personal opinion isn't important." It has suddenly become important, since you're implying that you disagree.

"Not exactly." Instead, say, "Here's the situation."

"That's not my department." Instead say, "This department is responsible for…"

"We're doing the best we can with the resources available." This begs the question "Are you doing enough?" Instead, say "We're doing this, this and this."

"At this point in time." The word is "now."

"But." Replace but with "and."

"Not." There are exceptions, but not usually.

"As I said before, or like I said earlier, or as I already told you." Instead, simply repeat the quote.

The power of repetition

Repetition is one of the most effective communication control techniques.

When you repeat a key pillar in an interview that will be fully edited, avoid the phrase "as I said before." If necessary, you may have to repeat your pillars or support statements several times and you must deliver the fifth repeat as freshly as the first. My method is to repeat your pillar, but change your sparkler or use a different bridge each time you repeat it.

If you don't think repetition is important, why do you suppose politicians stick to a script and repeat it all day? There's no value in intellectualizing over the use of repetition — it works. It's saved me more than any other technique, no matter how tough it felt to sound like a broken record.

Repetition reinforces the message. The "big lie" theory holds that if you say something loud enough, long enough and often enough, people will think it's true. It is a method used by big liars throughout history.

Repetition is one of the most effective communication control techniques.

Use repetition when you have too much media attention.

Reporters use repetition throughout a news story — which is, after all, a composition that follows a structure. The news media repeat major stories for hours throughout the day. The same headlined TV messages are repeated again and again. News is highly repetitive. Your message must often be, too.

The exceptions to the repetitive approach?

- ☞ When you're not at first successful with your message, improve it.
- ☞ When you're trying to get media attention and early messages aren't working.
- ☞ When you're starting a campaign to develop long-term relationships with a small number of specific but unique types of reporters.

Repetition is one of the most effective communication control techniques.

Helpful phrases to slow down the charging reporter

We call these statements re-asks, because they require the reporter to repeat the question while you engage your brain for the answer. They can be very helpful on the witness stand, until the judge orders you to answer. There's no judge in a media interview, except the final viewer.

"Could you be a bit more specific, please."

"I'm sorry, what was the first part of your question?" (For use with multiple questions.)

"Could you please re-phrase the question?"

Here are some last-resort, "I"-word answers to buy time, for use in extreme situations only:

"I would be interested in knowing how you arrived at that conclusion."

"I'm not sure I understand how that question relates to my function."

"I really have nothing else to add to what I've already told you."

"I'm sure if you check my earlier answers, you'll see that I addressed that issue."

"I'm sorry if I haven't explained this very well." I use this when I want to let the reporter know I think they're stupid. After saying this, I repeat what I've already said, but more slowly.

Handling the ambush allegation

If you've seen CBS's *60 Minutes,* you'll know the situation. The question comes from out of the blue and you're entirely unprepared. The question is like this one:

> **Q.** *"Is it true some of your inspectors have been taking free gifts from Company A?"*
> **A.** *"I'm very sorry. You've raised a question that I have no information about. Would you object if I called in Marge Taylor from the office next door to sit in on this interview and take down some information so that we can look into the situation. (Later, when Marge arrives:) What, specifically, is it that you're alleging? Can you describe the persons involved? When is this alleged to have happened?"*

These are situations where you don't know what the reporter knows. You can't confirm or deny something you've just heard. Ask the reporter for some facts. If there has been wrongdoing, get your organization on the right side of the issue quickly. Become part of the solution rather than part of the perceived problem.

If you start to look guilty, people will think you are.

You might even ask the reporter not to divulge certain information so as not to jeopardize your own internal investigation. In return, the reporter might ask for exclusive rights to first information, something which is entirely appropriate. You now have a partner in the reporter, however unlikely the partnership.

Be very careful about asking the reporter where they got their information. Questions like, "Who told you that?" only illustrate your weakness. Reporters will likely hide behind their right to protect their sources. The same reporters are also apt to ask you the same kinds of questions, or variations like:

"Have you been told you can't talk about this?"

"Who told you that you couldn't talk about this?"

"Are you covering up something?"

The pause

In an edited interview, we seldom hear the question. Check out the newspaper — how do we know what the question was that generated the quote? We don't, unless the quote contains the word "not."

When I become the reporter from hell in our training classes, a majority of my students can hardly wait to start answering the first bombshell question, sometimes even before I can get the microphone over to them.

In a live-to-air interview, reporters hate silence — they know that within a few seconds they can lose half their audience. You can master the use of silence, too, especially in an edited interview. Always follow silence with planned responses, even if you use the silence to compose the material.

Think of all the times in your life when you shouldn't have said the first thing that popped into your head. If you need 10 or 15 seconds to develop your answer, take it.

The pause won't make the question go away — you must answer with a statement from your media line. Looking pensive or insightful can be part of your strategy, and saying nothing can be the best answer — it's better than saying "No comment."

In over 30 years of granting media interviews, I've only had a pause left in the news story once, and that was for dramatic effect. It worked out fine.

Chapter Eleven

Sparklers — Displaying your excellence

Any public relations mouthpiece can step up to a microphone and say that safety is our number one concern. It takes a subject expert to follow that up and say, "Let me tell you about a new safety program we're introducing. It's called..."

Any news story will contain statements that define the issue — whether attributed to you or not. Good news stories and hence, good media lines, contain examples — sparklers that help reporters more than any other statement.

When it comes to professional spokespersons, sparklers separate the wheat from the chaff.

Why sparklers pay off

You can't illustrate your pillars without sparklers — they illuminate your message. Sparklers serve several essential, strategic functions:

- ☞ They provide the spokesperson with credibility by allowing you to display your knowledge in an interesting, informative and entertaining way.

- ☞ They give reporters options to develop leads for their stories and you can use different sparklers with different reporters, while using the same pillars with everyone.

- ☞ Sparklers are the most strategic part of your media strategy — they serve as proof statements to your pillars.

- ☞ They illustrate your MBWA skills — management by walking around.

- ☞ Sparklers appeal to the natural sense of curiosity that should exist in all reporters.

- ☞ Sparklers are the most effective way of talking about your pet project.

- ☞ Sparklers can put a human face on the topic.

- ☞ They can vary upwards from a few words, allowing you to fill time. If you have an hour to fill, you'll want lots of sparklers.

- ☞ Sparklers work best when they're volunteered early in the interview, without the reporter having to ask you for this information.

- ☞ There are lots of different types of sparklers. **The success-story sparkler** is an example of what you're doing or what you're achieving. Here's where you do your bragging.

☞ Sparklers work best with bridging phrases like, "Let me give you an example" or "For instance." The more sparklers in your media line, the better equipped the spokesperson.

Rattle off some numbers

The statistical sparkler allows you to use verifiable, provable data in an interesting way. Your use of such information is only limited by your memory and several of our students have displayed tremendous recall powers to great success. You cannot, however, ignore negative data. Nor can you ever expect to deceive reporters whose skill for figures may exceed yours.

Percentages are vital in most statistical situations, but always be sure that you're not comparing apples to oranges, as the saying goes.

Statistical analogies can be very important, and reporters use them all the time. In a recent feature on highway safety, the *Toronto Star* pointed out that the number of people killed in traffic accidents in Ontario in one year was equal to the population of the village of Manotick.

A government safety expert should be very reluctant to use that specific analogy for fear of offending, upsetting or angering the good people in Manotick, but they might make the comparison based on the population of a "medium-sized village in many parts of the province."

The reporter's story went on to question what the public's (and presumably, the media's) reaction would be if 1,000 people had been killed in an explosion that was preventable and what kind of enquiry there might be into the accident.

Use teaching aids

Sparklers provide detail that adds to your pillars. In a news release, sparklers are the backgrounders you attach to the one-page release. In this book, case studies are sparklers.

Lists of highlights work well. The experts suggest lists of 7, 10 or 12. Add charts, graphs, photos, diagrams.

A photo opportunity is a visual sparkler created to give photographers subjects for pictures — video or print. Just ask Greenpeace about "media opportunities," sometimes called stunts by their critics. Every day organizations create events just for the news media — these are "sparklers."

Anecdotal sparklers

The anecdotal sparkler is the mainstay of sales pitches by vacuum-cleaner sellers, snake-oil peddlers and door-to-door marketing people.

If it's true, if you know in your heart it's right and if you can say it in good conscience, there are benefits from a message like, "One of our customers told us that what she likes about the new jewellery line is the price, the selection and the colors."

A large proportion of news stories are anecdotal. Recall the Ice Storm '98 stories about individuals affected by the storm. Each story was an anecdote, to symbolize the impact on people, just as the famous Willie Horton advertisement was an anecdotal example.

Case Study: Caring about agricultural land — Farmer Brown's daughter

You may be a highway engineer involved in land expropriation. The news says that poor Farmer Brown will have to give up half his orchard for a new highway. His daughter won't be able to go to nursing school because of his loss of income.

> *There is not one thing you can do to help his daughter, short of paying her tuition out of your pocket. What you can do, however, is develop sparklers that illustrate the time and effort you put into examining alternatives. Explain your commitment to preserving agricultural land, and illustrate what the options would have cost, without making Farmer Brown look greedy, of course.*
>
> *"Expropriating land is a last resort. Yes, we're always concerned about acquiring fruitlands for highway use. That's why we did this, this and this. That's why we worked so closely with agencies like this one, this one and this one. That's why we looked so closely at alternatives that are safe, efficient and environmentally responsible. Our main concern, of course, is safety.*
>
> *"Let me give you an example of how we study land use. First, we..." You have now bridged to your next sparkler.*

Unique features sparklers

The unique features sparkler allows the spokesperson the greatest opportunity to volunteer information. There are few, if any, limits on the length of these sparklers because they are self-contained and the reporter will be aware that you are making a certain point. Use a bridging phrase like:

> "One of the unique features of this product is..."
>
> "What makes this product so popular is the..."
>
> "Some of the things we found in our study surprised us, for instance..."
>
> "This is the first widget to include a new safety feature called..."
>
> "We didn't expect the reaction we got when we asked store-owners to display this product. We found that not only did..."

The sparkler as parable

You'll find several parables throughout this book — wherever you see a reference to a case study, that's an experiential parable. I often use them in class when the students start to get restless or tired. They perk up the training, and the instructor. I can't teach without them.

Experiential communication is an essential teaching method used by all of the world's greatest communicators. It's really a unique blend of story-telling with teaching and with giving a speech.

If you know that it takes about four minutes to tell the story of your first day in the new office and what you learned that day, you have an invaluable time-filler that also helps to deliver lessons, messages or teaching points for today's situation.

One day in class, one of my students pointed out, "I can tell when you're going to deliver a sparkler to us. Your body language changes, the relaxation is noticeable and you capture our attention. And you've said, at least four times so far, 'That reminds me of the time at Pearson Airport when…'" I love to tell airport stories.

Why do I use so many airport experiences in my training? Because they make a point, students can relate to airport experiences, and they serve to provide me with some degree of credibility, or at least I hope they do. Almost everyone flies, although few enjoy the experience any more.

In developing a speech or presentation, you only have to indicate on your cue cards "insert story about…" and you've got a packaged message that you've probably told dozens of times.

You've test-marketed this story, you know how it flows and your mind isn't challenged to make up a new message. And since the story went over well in Phoenix, it probably will go over well in Tucson.

The "chapter and verse" sparkler

It can be necessary to quote certain documents or regulations exactly as they're worded in legislation or in a report:

> "Section five, sub-section two, paragraph three of the Clean Driveway Act states, 'No driveway shall be built…'"

> "The judge's ruling was clear when she said, and I quote, 'The parties are required, by the end of April, 2005, to carry out…'"

> "Our shipping instructions are quite clear. They state, on page one, 'Always wear protective eye-covering before assembling this toy.'"

> "Our policy on late-arrivals is printed clearly on the back of each ticket and at each entrance to the hall. 'No one will be seated once the program starts.' We regret the inconvenience, but this policy was developed to serve our customers who arrive on time and don't wish to be disturbed. It also means that the actors will not be disrupted in their performance. Let me give you an example of what happened one time…"

The "for instance"

The most successful sparklers are "for instances" that illustrate your professionalism, knowledge and expertise. You can never have too many for an interview. There are no strict limits on their lengths. Reporters love them. They generate better news stories.

In January, 1996, *Toronto Star* reporter Antonia Zerbisias sat in on one of my courses and later wrote a feature article for the Insight page. The article, entitled *"The ABC's of becoming media savvy"* appeared on May 27, 1996, your author's 46th birthday.

In the course of her writing, she called me back several times as she focused in on how she wanted to present the topic. In the fourth interview, which I later analyzed, I discovered that to every one of her questions, I used a sparkler for an answer.

She asked a question like, "Aren't you afraid that you're simply trivializing the whole interview process?"

The answer was something like: "Let me tell you about some of the work we did with a client who worked in the area of... The interview was for the evening news, so we knew that the reporter needed one or two 10-second quotes. That's why we concentrated on... The result was..."

The next question was "But aren't you just stage-managing the news?"

The answer was something like: "Let's look at what's happening in the typical Toronto TV newsroom. Each reporter is assigned a story of about 90 seconds in length and given a short time to arrange a photographer, book the interview appointment and grab one or two quotes. The spokespersons we train are well-equipped to provide reporters with exactly what they need. For instance, one of our students, who was being interviewed about..."

Here are some examples of sparkler statements that various spokespersons will want to be able to volunteer in an interview — because they work best when they're volunteered, without the reporter having to dig them out of you.

For the police department spokesperson being interviewed about home security:

> "There are several things that people can do to make their homes more secure. Start by..."
>
> "Members of the Westhaven Police Force have conducted over 200 information sessions on home security to over 4,000 people since 2002. We focus on three main areas..."
>
> "One of the questions we're most often asked is about illumination. We recommend..."

For the police department drug education officer:

> "One of the biggest factors in teen drug use is peer pressure. That's why we've introduced a program called…"

> "Let me give you an example of some of the drugs recently discovered in a teenager's room by her parents. They included…"

For the United Way spokesperson, your sparklers can include a wide range of heart-warming examples of the fine work being carried out by your member agencies. The challenge is to choose agencies that best meet the goals of the United Way and that are not involved in other controversy, keeping in mind that you're not a spokesperson for those agencies.

Many United Way advertising campaigns consist of photos of people who are examples (sparklers) of how the agency helps. You can use them and probably should use them in an interview.

You might choose generic examples of some of the wide range of services provided to seniors, youth or people with disabilities. You might equip yourself with personal testimonials from people who donate to the United Way. Use the stories to get people to donate.

The Westhaven Hospital spokesperson should be equipped with a number of specific examples of where cost-savings have been achieved, such as in laundry services, food preparation, computerized administration changes and housekeeping. Keep in mind that many of these may be controversial and that they may have been "very tough decisions."

The spokesperson for Westhaven Widgets should be equipped with sparklers about the unique manufacturing improvements that have been made, or how your widgets are "longer-lasting, due to a unique bonding technology developed right here in Westhaven. One of our staff came up with the suggestion that we…" This type of example boosts employee morale.

The spokesperson for Westhaven Restaurant Systems should be equipped with sparklers that address high standards for purchasing food products. Example:

> "When we purchase produce, we insist on a flavor test by our senior chef, Tony Baloney. In season, we only serve tomatoes grown locally by our own Westhaven Greenhouses because they've proven to be the most popular with our customers, especially in our Papa Tony's Pasta selections. We buy all our cheese from…"

The key to using sparklers is to choose them in advance and volunteer them without being asked. Reporters love them and many news stories start with sparklers chosen by the reporter as good examples of the story they want to tell. Sparklers can range from a few words to several thousand.

You won't learn about sparklers if you never get out from behind your desk and learn to manage by walking around. Your MBWA skills will separate you from the pack and position your public-spirited professionalism more effectively than any other media message.

When is it OK to cry?

What makes this issue important to you personally? Many of us have a personal stake in the issue — a special anecdote or situation that symbolizes why we care. At some stage, a reporter may push the envelope and you must be ready. Or, you may be speaking to a public meeting when your professionalism (and your parentage) is challenged.

For photo radar, it's the police officer who says that the worst part of their job is telling a family that someone has died in a traffic accident that was preventable. It's the engineer whose husband and children use a highway to travel every day. It's the drug education officer who has a friend with an addiction. It's the hospital administrator who's seen first hand how new surgical techniques can save lives.

Will it make you cry to tell it? Then it must be powerfully important to you, to the reporter and to the final receiver of the message.

The sermon sparklers

I've taught a number of church leaders, many of them from the Metropolitan Community Churches, based in West Hollywood, California.

MCC Founder and Moderator, the Rev. Elder Dr. Troy D. Perry is a friend of mine, and a client. He remains a champion for human rights in America and has been a guest at the White House several times. I was best man at his wedding July 16, 2003 in Toronto.

Having heard him preach and/or speak over a hundred times, let me share what I believe is his sermon formula, adapted to pillars, supports and sparklers:

☞ **Sermon Title** — a support statement

☞ **Opening Scripture** — support

☞ **Warm-up support phrase** — "If you love…"

☞ **Sermon Theme** — defining pillar. "Today, I want to talk about…"

☞ **Sermon outline** — three explaining or describing pillars:

Pillar one with up to three sub-pillars

· Supports and one or more sparklers of one to three minutes each.

Pillar two.

· Same.

Pillar three.

· Same.

☞ **Summary statements**

☞ **Sprinkle through with one-liners and quotes.**

Rev. Perry can put together a sermon on almost any topic with his stable of sparklers (he has hundreds of stories he can insert, all guaranteed to make his points).

All he has to remember is, after pillar one, tell the dead dog story. After pillar two, tell the story about meeting Bishop Desmond Tutu in South Africa. After pillar three, talk about the new fund-raising campaign.

If he has time, he talks about his latest visit to the White House. Or when Rev. Nancy Wilson shaved her legs to meet the Pope.

Context, background and history

Here are some areas to consider in preparing a background briefing:

- **Political decisions and activity**
- **New research or study**
- **Issues raised by area residents during public meetings**
- **Lessons learned from experience of others**
- **The changing roles of various groups and organizations over time**
- **Changing regulatory activity**
- **Changes in leadership style or personality type**
- **Changes in funding levels or sources**
- **New developments in technology**
- **Market forces at work, changing buying habits or sales trends**
- **New competitive situations or purchasing habits**
- **Environmental impacts, legislation or practices**
- **Changes in public opinion**

The opportunities to develop background and proper context are endless. Each of these items might be 100 words — or they might fill an encyclopedia. **You'll want to match the amount of material to meet the interview time.**

A reputation for story-telling

History fascinates certain audiences. The history of your organization. The first or most noteworthy general manager or leader. The history of a facility or service. The history of an issue. The first store or factory and what the situation was like. The first customer. The first vehicle to pass through.

That small widget-maker who once hauled his product by horse and wagon and today owns a theme park in Bavaria. What it was like to be a pilot before computers and what it's like today.

The effect of a war, or another major event, on your organization. What your organization was like on the day that President Kennedy was assassinated.

What it was like to build things when you first started. How files were processed in 1996 and how they're handled now. How have things improved, or not improved, over the years?

What people earned then and what they could buy then compared to the situation now. How decisions were made then and how they're made now. Who made the decisions and how long it took to make them.

Blessed are the story-tellers, for they shall be heard.

Often, the quality of history is determined by the quality of its telling. Certain writers have the gift to take mundane matters and bring them to life through the richness of their language, their choice of words, their speaking skills.

Case Study: Caring about history — Open line opportunity

An airport manager, friend and client, George Elliott, was once the guest on an open-line show when suddenly it was discovered that the person who was to follow him had canceled out. Host Peter Warren asked George to stay for the next hour, even though they agreed their first topic had been exhausted.

They invited callers to telephone with any interesting stories about Winnipeg International Airport — their first flight, family joys or tragedies where the airport played a part. People who lived near the airport, retired airport, airline or other employees were encouraged to call with their stories. They invited people to share their sparklers.

The lines lit up immediately. The host and the guest sat back and allowed people to tell their stories and a few of the callers were contacted later and invited to serve on the airport's 60th anniversary committee.

Some people later donated historical photos for the airport's anniversary book. Others volunteered to act as hosts for the airport's anniversary open house. The open-line show turned into a community relations success. The anniversary sparkled with community involvement. Open line shows aren't always that enjoyable, but history seems to touch people's hearts in mysterious ways. Told well, history is vital to a media spokesperson's package.

When I left my job in Winnipeg in 1980, at my farewell party I received one of the highest compliments that any PR person could receive. A CBC news director told me that I had succeeded in making George Elliott the most accessible public servant in the city.

MEDIA LINES — A summary

Your media line may not need to follow a strict format in the actual interview. Generally you'll want to deliver your pillars early and you combine a pillar with supports and sparklers, as required.

As a spokesperson, you'll be required to write your own core statements, use those provided you by your organization, or work with writers who specialize in the art of ***MediaSpeak***.

Since the reporter is often seeking only one or two quotes, the ones selected are likely to meet all the reporter's requirements — they're short, concise and interesting.

Start to observe news media coverage in a new way and notice how often the favorable quotes we see are examples of pillars, supports or sparklers.

Start observing how successful politicians communicate in core statements, often written in short sentences using three's.

There's no shortage of examples of spokespersons who make mistakes — we see the results in headlines and story leads all the time. Often, the spokesperson knows that the message could have been better had more care and thought been taken to handle a certain issue.

A slip of the tongue is all it takes to damage a program, destroy relationships with public interest groups or provide your critics with the ammunition they need to attack you.

It's our goal to place you in a position where you never have to say "If only I had said it this way instead…" And you'll never have to say "No comment."

A Spin Doctor's Lament:

If I have a fancy title, a great salary and a good track record, I am only as good as my next media interview.

If I have a wealth of training and experience, my credibility must be established in each and every media encounter and my success depends upon "spin."

Spin involves research, strategy and presentation. It involves influence, clout and delivery.

Spin is short, concise and often direct and is aimed at the recipient of the message, not the messenger.

Spin is about angle, approach, wording, style, emotion, timing and choice of medium.

Spin doctors very seldom get credit for their skills, but when things go wrong, they always get the blame.

Spin is not about declaring one's excellence — it's about displaying it.

Spin is not about stating one's greatness, it's about proving it.

Spin is based on truth, tact and timing.

Spin is not about saying what you think. It's about thinking about what you say.

Spin is not saying what's on your mind, but being mindful of what you say.

Spin should never boast, ridicule or belittle, but should allow your presentation of the facts to speak for itself.

When I was an amateur, I spoke like an amateur, but even when I became a professional, I remained a student and never stopped learning from the successes or failures of myself and others.

Appendix

Write Better News Releases:
How to improve your "Pick-Up Rate"

by Ian Taylor
President
Never Say "NO COMMENT" Inc.

Pick-up rate?

We're not talking about your success in a singles' bar here. We're talking about the success of your news releases and related publicity materials. Are they getting "picked up" and used by newsrooms? Are they creating news? Do you want them to?

When you want your news release to generate news, it's a lot like going fishing. You have to know what the fish are looking for, how to hook them and how to land them on any given news day.

When we conduct a communications audit with a client, we often start by looking at the success of their communications efforts, as reflected by the success of their news releases.

A news release can be the cheapest way to get your message on the front page, or, when poorly written, an expensive way to be ignored by the world.

The kinds of questions we'll ask:

- ☞ **Are news editors or newsrooms using your news releases to generate stories or do they end up generating nothing after weeks of work and high staff costs?**

- ☞ **What kind of reputation do you and your news releases have in newsrooms?**

- ☞ **How close is the resulting headline to your news release headline?**

- ☞ **When your releases are used, are editors or reporters changing the lead sentence or using your lead in their story?**

- ☞ **What lead are they using?**

- ☞ **How closely does the story content and structure match the original news release?**

- ☞ **Does the final news story meet your communications objectives and why or why not?**

- ☞ **Are you prepared to change the way you communicate, starting by re-examining the way you write news releases?**

If your news releases are failing, chances are you haven't written them to read and sound like a news story.

Chances are your present news releases sound like a PR promotion piece or paid advertising posing as news.

Or else they're so bureaucratic that the reader can't easily understand what you're saying without having to translate the copy into plain talk.

Learn to write your releases in *MediaSpeak,* not PR-Speak or Bureaucrat-Speak.

When you over-inflate the positive language, or position profits and egos above the public interest, your news releases will reflect that.

They will never get past a junior newsroom clerk.

If they are used, they'll be entirely re-written by responsible editors or re-write staff.

When you write in *MediaSpeak,* your writing style matches news style.

What's the secret? Your core news release message must be positioned in the public interest; it must be written in plain talk; and it must reflect your organization's professionalism.

The result: little editing is required by newsrooms in order to create a final news story because you've applied news style to your news release.

In short, a broadcast reporter can "rip and read" your release.

This term originated in the days when a news release arrived on a roll of teletype paper and a radio reporter could simply rip it off the machine and read it on air without editing it. If a reporter can do that with your news release, you've written it in *MediaSpeak.*

Tips:

- ☞ **Read your news release out loud, like a radio reporter would.**
- ☞ **Re-write it until you make it sound like a real news story.**
- ☞ **Always write in the third person.**
- ☞ **Target a grade 8, or lower, comprehension level.**
- ☞ **Keep the writing tight, fact-filled and interesting.**
- ☞ **Write short sentences and keep your paragraphs short, just like newspapers do. This results in more white space on the page and draws the reader into the words, not away from them.**

- ☞ **Notice how newspaper articles look on the page. See how short the paragraphs are.**

- ☞ **Count the words or check the timing of TV news stories. If the typical TV news item is 90 seconds, that's about 200 words. If your goal is to create a 90 second news story, write a 200 word news release and no more.**

- ☞ **Add extra material as detailed "backgrounders."**

Signs of News Release PR-Speak and how to avoid mistakes:

1. **Mistake: Leading the release with the name of the organization head**

Avoid: "Mayor Angela P. Daley III, PhD, LLB, B of B. announced a tax cut today…"

Instead: "Taxpayers will benefit from a new tax cut that will save over…"

Tip: Wait and place the announcer's name or organization name in the second or even third paragraph, devoting the first two paragraphs to the public interest and to key benefits, concerns or reasons.

2. **Mistake: Proof without pudding, fabulous without facts – declaring excellence without displaying it**

Avoid superlatives without proof. Instead of loudly declaring your excellence, convincingly display your excellence with product knowledge, sparkling examples and vivid imagery.

Avoid opinions.

Instead, use facts, comparisons, statistics to prove your case, provide you with credibility and position yourself as a professional.

3. **Mistakes: Watered down words, indirect or passive language, bureaucratized wording, buzzwords, non-committal, non-specific statements**

Write the way reporters write, unless you want to sound like a publicist or a bureaucrat.

Use the world's oldest and most effective power words like: **work, help, build, serve, improve, protect, save, safety, future, community.**

If you start making major improvements in your news releases, you'll soon start to improve your pick-up rate.

You'll enhance your reputation in newsrooms as credible sources and you'll more effectively influence public opinion with your messages. Make the changes today and see the results for yourself – on the front pages.

Ian Taylor is a former public affairs manager at Toronto's Pearson International airport, where his news releases had some of the highest pick-up rates in newsrooms because they were written in the public interest, in plain-talk, and with some degree of professionalism.

For help with your news releases, call Ian at 416-466-1560.

Ian Taylor

Ian Taylor spent the first half of his career overcoming a bureaucratic culture intent on reducing and restricting public communications.

He's spent the second half of his career training and advising clients to conduct themselves in the public interest, to talk like real people and to display their excellence without declaring it.

Ian is the former chief spokesperson and public affairs manager at Toronto's Lester B. Pearson International Airport. He has lectured at the International Aviation Management Training Institute in Montreal, serving aviation officials from over 100 countries.

In 2001, Ian was sub-contracted for work with the Port Authority of New York and New Jersey for their three international airports' World Class Customer Service Project. The work ended with the events of September 11, 2001.

Ian is a professional corporate speaker, trainer and consultant based in Toronto and Turkey Point, Ontario.